PROCEEDINGS OF 2018 EEC/WTERT CONFERENCE

Edited by

Marco J. Castaldi
The City College of New York, Earth Engineering Center|CCNY

Francesco Patuzzi
Free University of Bolzano

Simona Ciuta
RRT Design & Construction

Swanand Tupsakhare
Southern Research Institute

Demetra Tsiamis
Earth Engineering Center|CCNY

ASME PRESS

PREFACE

The 2018 bi-Annual EEC/WTERT conference was held this year at The City College of New York hosted by The Earth Engineering Center at City College (EEC|CCNY). The conference brought together the best minds in academia, industry, and government representing 12 countries across five continents to discuss important current topics in waste. The meeting provided an unprecedented and timely viewpoint of the global activities in sustainable waste management. Experienced professionals and next generation of engineers presented and worked together to develop a blueprint of the future of waste sustainability. The takeaway message from the conference was that the circular economy is coming. Good engineering will get us there, but until then we will need to strengthen collaboration efforts using all solutions available to properly manage the growing waste streams. Importantly it is the actual activities that are deployed, proven and used to shape decisions from regional to municipal levels.

This conference originated in 2004 and has enjoyed good visibility and international recognition. Invited speakers provide an updated perspective on sustainable waste management. This year's conference further broadened the scope of past bi-annual conferences to incorporate discussion of all waste management practices comprising the sustainable waste management hierarchy. It expanded its scope to include representation from government agencies, local municipalities, and forward thinking foundations. A number of presentations ranged from the circular economy to reuse, recycling, and composting programs, to the impact of communication on public acceptance of waste-to-energy. The remainder of conference had presentations and discussions on technical aspects and processes, perspectives from local municipalities, industry, students and regulatory personnel. The conference also included poster sessions that featured the innovative waste-related research conducted by student research associates of the EEC|CCNY. A summary manuscript from the 2016 CCNY conference, "Megacity Waste Management Dialogue: The NYC Roadmap for Waste as a Resource" leads off this special issue to set the stage and demonstrate the continuity of events.

The Conference was co-sponsored by the Grove School of Engineering with special contributions from the Chemical Engineering Department and the Earth System Science and Environmental Engineering Program. The ASME Material and Energy Recovery Division and ASME Research Committee on Energy, Environment, and Waste sponsored the Awards Dinner recognizing the student scholarship winners and the Pioneer Award given to Lew Clark. Industrial support was provided by HDR, GBB and Martin GmbH. We express our sincerest gratitude to the participants and to the reviewers of the manuscripts.

The manuscripts assembled here reflect the presentations and provide a summary account of what was presented during the two-day event. The accompanying presentations can be viewed at http://ccnyeec.org/2018-wtert-conference/.

Marco J. Castaldi, Director of EEC|CCNY and WTERT-USA   April 17, 2019

TABLE OF CONTENTS

ABOUT THE EDITORS

EDITOR

Professor Marco J. Castaldi holds a B.S. from Manhattan College and M.S./Ph.D. from UCLA (University of California at Los Angeles) all in Chemical Engineering with minors in Astrophysics and Advanced Theoretical Physics. He holds seven patents and two pending applications in the areas of catalysis and combustion. He has authored over 95 peer reviewed publications, three textbooks and 7 book chapters. He was the Editor of the North American Waste to Energy Conference (NAWTEC) Series (ISBN: 978-0-7918-4393-2) and currently serves as an Editorial Board Member of *Waste and Biomass Valorization* (ISSN: 1877-2641), *Catalysts* (ISSN 2073-4344), *Journal of Environmental Engineering* (ISSN: 0733-9372) and *Applied Energy* (ISSN: 0306-2619). He has worked in industry for 10 years and was an Associate Professor at Columbia's Earth & Environmental Engineering Department prior to coming to City College in the Fall 2012. He is a Fellow of ASME, AIChE, the Fulbright Commission and the National Academy of Engineering, Frontiers of Engineering Education.

ASSOCIATE EDITORS

Dr. Francesco Patuzzi holds a B.S. and a M.S. in Environmental Engineering from the University of Trento. He is currently Assistant Professor at the Faculty of Science and Technology at the Free University of Bozen-Bolzano (Italy), where he received his Ph.D. in 2014. His research activities are mainly related to the study of thermochemical conversion of ligno-cellulosic biomasses, with a particular focus on the characterization of pyrolysis and gasification processes, the valorization of their by-products and the analysis of the technologies for the separated or combined heat and power (CHP) production from biomass. He is and has been involved in several national and international research project dealing with these topics. The main investigated processes are gasification and pyrolysis, but also traditional processes such as combustion, as well as innovative valorization pathways such as hydrothermal carbonization are included in the research interests, together with possible syngas upgrading routes such as reforming, methanation and Fischer-Tropsch synthesis.

Dr. Simona Ciuta received her B.S. in Power Engineering from Politehnica University of Bucharest. Dr. Ciuta developed her bachelor's thesis at the University of Trento, Italy after being awarded an Erasmus Scholarship. She obtained a Master's Degree in Environmental Management and a Ph.D. in Power Engineering from Politehnica University of Bucharest, Romania. During her Ph.D. she was invited at University of Trento, Italy for research stages to further develop her thesis. Currently, Dr. Ciuta works as a Project Engineer for RRT Design & Construction. Prior, she worked as a Post-Doctoral Research Associate for the Combustion and Catalysis Laboratory in the Department of Chemical Engineering at City College of New York (CCNY). Dr. Ciuta`s research focused on waste materials conversion into energy, waste management, thermal processes such as gasification and pyrolysis. Dr. Ciuta developed a one of a kind intra-particle gas sampling technique that provides significant insight into fundamental reaction sequences happening during thermal decomposition of waste materials.

Dr. Swanand Tupsakhare serves as a Principal Investigator and Technical Lead on US Department of Energy Projects at Southern Research Institute. He has been involved in applied research in the areas of carbon capture/utilization and bio-based chemical synthesis to offer innovative and cost-effective solutions while reducing the burden on the environment (reduced waste, emissions & toxic chemicals use). Previously, he worked on investigating the elevated temperatures in Municipal Solid Waste (MSW) landfills. Dr. Tupsakhare received a Ph.D. in Chemical Engineering from the City University of New York and Bachelor's in Chemical Technology from Institute of Chemical Technology (ICT/UDCT), Mumbai, India.

Demetra Tsiamis is the Associate Director of the Earth Engineering Center at the City College of New York (EEC|CCNY). Ms. Tsiamis manages the center's applied research programs on waste sustainability sponsored by private industry and government agencies. She is the lead engineer of technical due diligences of waste gasification and pyrolysis technologies and has led research programs for Covanta, the American Chemistry Council, and the New York City Department of Sanitation. Ms. Tsiamis received a Master's in chemical engineering from Columbia University and a Bachelor's in chemical engineering from The Cooper Union with a minor in environmental engineering. Her Master's thesis at Columbia University under the co-advisement of Professors Nickolas Themelis and Marco Castaldi investigated the technical feasibility of commercial pyrolysis technologies in the treatment of New York City's non-recyclable plastics. Prior to EEC|CCNY, Ms. Tsiamis worked as a field engineer for Langan Engineering and Environmental Services in New York.

MEGACITY WASTE MANAGEMENT DIALOGUE: REVIEW OF WORKSHOP FINDINGS

Simona Ciuta, RRT Design & Construction

Demetra Tsiamis, Earth Engineering Center|CCNY

Nathiel Egosi, RRT Design & Construction

Marco J. Castaldi, The City College of New York, Earth Engineering Center|CCNY

ABSTRACT

The Earth Engineering Center (EEC) at The City College of New York (CCNY) hosted an interdisciplinary workshop on sustainable waste management practices with the objectives to facilitate discussion among all stakeholders and professionals and educate the public. The subject matter covered all processes and technologies available for waste management as well as perspectives from municipal, regulatory and policy personnel. The primary outcome was to advance the waste infrastructure of municipalities by bridging the knowledge gap between policy makers, industry, academia and the public. This report presents the findings that emerged from the workshop breakout sessions focusing on lessons learned and forward-thinking concepts while referring to real scale applications that use best practices related to waste management and associated technologies.

INTRODUCTION

Total worldwide municipal solid waste (MSW) generation is estimated at 1.3 billion tons per year according to the World Bank Group. The United States produces 625,000 tons per day, or about 4 pounds per person per day, which is nearly the highest in the world. For the past 40 years, there has been a major push throughout Europe and the United States to reduce energy recovery from waste from the waste streams remaining after reuse and recycle.

Even with all these efforts, a report from the World Bank released in June 2012 [1] revealed some concerning information. Annual MSW generation from the 161 countries analyzed in the report is expected to increase to 2.2 billion tons by 2025. In addition to that, waste generation rates were predicted in the report to be more than double in the next 20 years in 38 countries classified in the report as low-income countries. If those predications are accurate, the problems associated with MSW management will become the main issue related to human and environmental health.

Waste pollution is an issue that all communities worldwide must manage, regardless of geography, culture, or economic standing. Improper waste management can lead to disease, poor standard of living, and depletion of natural resources. Nearly 1/3 of the food generated worldwide is wasted. Approximately 90% of third world countries manage their waste through unsanitary landfills and open burning, which contaminates water supply and produces harmful air emissions. In 2015, about 80% of the waste worldwide was landfilled, which has been proven to be unsustainable [1]. Municipalities and local authorities aim to manage their communities' waste but they are limited in their efforts when the waste generator does not participate in the established programs. For example, the City of New York's Department of Sanitation has been operating a 30-year old curbside recycling program to divert recyclables from landfills. However, in 2016, only 17% of recyclable items were actually recycled by waste generators; the remaining 83% of recyclables ended up in a landfill [2]. Although mismanagement of waste may be attributed to human behavior, it is also largely attributed to a lack of education among communities and decision-makers. Communities are uneducated on the details of their waste diversion programs, such as recycling, composting, and reuse, and therefore dispose of many items that could otherwise be diverted from landfills.

Furthermore, communities are either unaware or uninformed of additional waste management technologies like anaerobic digestion and thermal treatment that can recover energy value from their waste. A recent survey conducted by the Earth Engineering Center (EEC) at the City College of New York (CCNY) revealed that 34% of New York residents did not know what waste to-energy was and when they were informed about its value, 88% wanted their waste to be managed incorporating this method [3]. Local authorities aim to assess waste management techniques and determine a best-practices approach for their communities based on available resources and current infrastructure. However, there is a lack of education at the level of policy-makers of up-to-date technical data and research in the area of sustainable waste management.

Landfilling is currently the dominant waste management practice in the US. Landfilling the majority of municipal solid waste (MSW) is not sustainable both economically and environmentally and has been determined to be the least desirable option for sustainable waste management [4]–[6]. It is viewed as the interim and safe approach as better methods are implemented and become ubiquitous in society. As a result, there is a pressing need for municipalities to develop alternative waste management practices that meet the long-term demands of increasing waste generation. Waste management is an issue that was recently highlighted as a major focus point on the agenda of New York City policy makers, as part of the One New York program (http://www1.nyc.gov/html/onenyc/index.html), the blueprint for long-term city infrastructure and environmental planning, to "reduce all of the city's waste by 90% by 2030- sending none of it out of state to be buried in landfills" (currently the City transports 100% of its MSW to out-of-city landfills and waste to energy facilities).

There are several key participants who are involved in addressing the issue of waste pollution – the industry, the regulators, the decision-makers, and the waste generators. The industry aims to provide a solution by developing technologies and materials that would reduce the negative environmental impact of waste. The regulators strive to use sound engineering and scientific principles to limit environmental impacts related to waste management. The decisions-makers aim to assess best waste practices for their community based on available resources and infrastructure. The waste generator is one of the main participants who needs to be incorporated into the discussions and activities on sustainable waste management and this can be done through education and engagement. Educational programs have been implemented sparsely throughout academic curriculums but the effect has not reached its full potential because education requires an integrated convergent approach that brings together all invested entities so that effective waste management solutions can be identified and implemented. Furthermore, waste management is not a "one size fits all" solution; it must be customized to the community because waste management issues vary based on geography, affluence, and social culture. For example, in many countries in Asia, South America and Africa, a waste management solution needs to be designed for a waste stream with high food waste content with a minimal waste collection and treatment infrastructure. Meanwhile in New York City, a waste management solution needs to be able to handle high concentrations of packaging wastes such as plastics and contend with multi-residential collection. Therefore, education and stakeholder awareness and engagement is critical in helping communities develop waste management solutions that are best suited to their situation.

MEGACITY WASTE MANAGEMENT DIALOGUE WORKSHOP

Waste management spans across several disciplines, from policy to engineering, and it has a direct impact on our communities. Discussions around waste management are currently very polarizing ranging from the point of view that greater efforts need to be made in the reduction of waste at the front end to ignorance of the fate and transport of waste when it is disposed.

Based on the US EPA's annual report, in 2013, over 87 million tons of MSW (i.e. 34.3%) were recycled and composted, thereby reducing GHG emissions by 186 million tons of CO_2eq [7].

However, programs that achieve the highest reported recycling from paper and plastic waste streams are 85% and 73%, respectively [8]. The remaining MSW, ranging from 15% in best case scenario to 27% based on national average, is technically not feasible to recycle, consequently resulting in an impossibility to achieve zero waste to landfill from recycling alone [9], [10]. However, energy can be extracted and can be used to enable processes that efficiently recover materials (i.e. Al, Cu, Fe) to be reused and incorporated into the material cycle. This message must be fully disseminated and understood.

EEC developed the Megacity Waste Management Dialogue to ensure that the right voices are heard in this debate with the end goal of finding a solution together. EEC hosted a workshop at the City College of New York (CCNY) on sustainable waste management practices as part of the effort to educate and increase the awareness of the public about waste infrastructure in the US and its current challenges, research and innovation in waste to energy, and sustainable waste practices employed around the world.

The waste industry consists primarily of chemical, mechanical, civil, and environmental engineers but the EEC workshop broadened the interactions to incorporate additional disciplines in engineering, policy, journalism, and architecture. Participants of the EEC workshop included the Commissioner of The New York City Department of Sanitation (DSNY), the Chief of Solid Waste Management of the United States Environmental Protection Agency (US EPA) Region 2, a journalist from Waste Dive, a researcher from the Environmental Research and Education Foundation (EREF) as well as engineers and consultants from all waste industries including recycling, waste-to-energy, and landfill.

The purpose of the workshop was to create a more visible dialogue with the end goal to close the gap between different disciplines in an effort to improve the sustainability of waste infrastructure in the US.

The workshop spanned two days and included presentations on the waste practices representing the waste management hierarchy, a global update of sustainable waste practices and the future of thermal conversion of waste, waste management in megacities around the world, human health risk assessment and environmental impact of all waste practices, research and development update from academia and the waste industry, the impact of policy and the market on the future of waste, and public opinion on overall waste sustainability. The participants at the workshop divided into 6 breakout sessions over the span of the 2 days to discuss the main themes of the presentations listed above.

The benefits of proper waste management are well known to experts in the field, however this report aims to present the assembled findings from the workshop which are valuable to a wide range of professionals from various disciplines involved in waste management and sustainability. This report also summarizes the recommendations from the workshop, while making connections with applications that are currently deployed and have been demonstrated to be effective.

WORKSHOP FINDINGS

During the different breakout sessions of the workshop, a series of issues and challenges were discussed. The main findings and recommendations emerged were grouped into the following categories and further detailed in the section below. Great importance was given to correlations of lessons learned from real case studies and hands-on field experiences of professionals working in this sector as engineers, consultants, and policy makers.

Academia/Research role in waste management

A major theme that was discussed was the role of academia and education in advancing society towards achieving its zero waste goals.

The inherent character of academic research fosters creativity and collaboration that gives rise to innovative and forward-looking solutions to the challenges that industry faces today. Research has an educating role not only for the public, but also for different industries, in particular proving the technical, less known details on waste paths from generation to disposal or treatment technologies. Research helps identify the waste streams that need more effective management or are not managed at all [11]–[14] and methods to manage them. In addition, the research community provides technical support to developers designing economically feasible solutions for advancing current and future technologies. There will always be room for research on how to advance and make processes more effective. Furthermore, it has been demonstrated that the research at EEC has mitigated potentially harmful environmental impacts of industries in the waste management sector. One example is the EEC research on elevated temperatures in landfills in which the findings guided landfills on the threshold conditions they should operate in to avoid issues that can arise during landfilling of wastes.

Research can also influence the public perception by addressing the concerns of the communities and policymakers based on technical data, knowledge, and experience. The participants in the workshop concluded that research is an important tool when regulations, guidelines and laws that are brought to congress need to be developed or improved. One example in which academia played an important role in this is EEC's research on fabric filter bags from air pollution baghouses of waste-to-energy plants. Local government and the public were concerned that a waste-to-energy plant's emissions were not meeting EPA criteria when the plant processed spent filters from the baghouse of its air pollution control system. Field data collected and analyzed by the EEC calculated that emissions increased by less than 0.2% when processing spent baghouse filters. The difference in emissions was immeasurable within the analytical detection limits of the typical monitoring systems for in-stack emissions therefore, it was concluded that combustion of spent baghouse filters does not significantly impact emissions of the combustor.

In addition to research, academia plays a key role in achieving waste sustainability because of education. All participants of the workshop agreed that education is necessary in order to achieve a zero waste future. This includes not only education of industries on their current waste footprint and how they can re-think their business model to achieve more circularity, but also, education and awareness of the public on their individual waste impact. In the recent EEC survey in New York City, 30.9 % of participants did not know where their trash went after they disposed of it. It is important that the public understands the impact of their waste habits and be technically educated on the waste practices in order for policy makers and engineers to effectively implement sustainable waste infrastructures.

Policies and regulations

A common finding that emerged from the different breakout sessions was that the lack of a consistent and comprehensive national waste policy in the United States (US) is a major obstacle in achieving improved waste sustainability in the US. There was a popular sentiment expressed during this conference that many challenges exist for defining and improving waste diversion in the US; fundamentally elected officials focus on not disrupting the citizens and undermining their support of current government. Attendees of the workshops expressed that the existing policies in the US need to be more frequently updated wherever possible.

One major impediment in forwarding an effective waste management in megacities is that waste handling and treating industries cannot influence the city, municipality, federal government or even consumers. Many times, the waste management industry is left with solving the problems created by others without having input in the development of regulations or incentive programs.

Policies and regulations should be based on more effective models, however there is not enough field data to validate the models yet. This is where research can fill the gap by gathering verifiable and consistent data, processing the information and developing working models to suit different situations [15], [16]. A sentiment that was shared by the EPA was that a lack of national policy also impacts national solid waste data collection and analyses. There is no regulation that requires reporting of data from waste treatment facilities (except for landfills). For any data that is available, there is no standardization amongst states of measurement and reporting of data in a homogeneous form that EPA can analyze and interpret with confidence. Therefore, the current methodology that is employed by EPA is based on industry reports and has proven to result in inaccuracies due to the inherent shortcomings of the methodology itself.

The federal policies that require compiling state reported quantities of MSW generated has proven to be inaccurate and inconsistent due to the differing approaches each state adopts for data collection and report. For example, in some states, MSW amounts generated or recycled are not fully recorded across all waste streams. This makes it very difficult to confidently develop or implement new regulations when the nature of the problem is not clearly known.

EEC made a recent comparison of EPA waste data with national waste data collected via independent surveys conducted by EREF and WtERT-US based on site-specific methodology. In this comparison, EREF and WtERT-US reported MSW generation in the US that was 43% and 31%, respectively, greater than that of EPA's reported tonnages. Furthermore, EREF and WtERT-US reported landfilling tonnages that were 61% and 49% higher than that of EPA reported tonnages, respectively.

In developing local and national policies, the US could follow the example of the EU which obligates its member states to develop waste management plans that are required to be reviewed and updated at least every six years. The landfill tax introduced by the EU government is another example of national policy that has significantly influenced the diversion of large volumes of waste from landfills in the EU and consequently incentivized the fast development and deployment of alternative waste management practices in Europe [17].

The EU also passed broad regulations that limit the amount of organics that will be sent to landfills. Also, many EU members are limiting landfilling only to inert materials, thus organizing municipalities to approach their waste management practices more sustainably.

Economic challenges

As expected, one major focus of the workshop was the issue of high capital and operating costs of waste management collection practice and processing technologies. The cost to dispose of waste is usually on the general population since they generate the waste, whereas the cost to process and treat the waste are not as clearly paid for by the consumers/waste generators. Limited funding is provided by government to design and build waste management facilities, as often these will not cover the costs for operations and maintenance. Therefore, funding to sustainable waste management technologies are expected to come from the revenues from sale of by-products such as energy and recyclables with insufficient user fees to fund the infrastructure to meet the stated policies and goals of the community.

Landfills being the least expensive option to manage waste, set a significant barrier to implement sound material and energy recovery solutions such as waste to energy (WTE) that requires high upfront costs followed by significant maintenance and operations costs. However, there are ways to offset some of these costs, for example through generation of electricity and district heating which can bring the cost of treating waste below the cost of landfill in some instances. For example, the City of Huntsville, Alabama owns a WTE facility that produces steam for the US Army's Redstone Arsenal resulting in a highly sustainable solution to not only manage solid waste but also to provide the heating and cooling needs for an entire community at a very low cost for both the City residents and the US Army.

In the workshop, it was brought up by participants from the thermal conversion technology sector of the waste management industry that it is difficult to economically compete with landfills in the United States. In addition, it was shared that the available technologies currently offer low flexibility because of high costs associated with alternative solutions. New technologies are very costly and only make sense in large scale applications. Anaerobic digestion of organics is gaining interest in particular since it targets the putrescible fraction of solid waste known for producing greenhouse gases when landfilled. However, industry is struggling with establishing the infrastructure to collect this fraction separately for introduction to the technology when communities realize how expensive it is compared to landfilling. Gasification and pyrolysis technologies are alternatives to WTE and can support energy recovery and zero waste goals however, little success has occurred on a commercial level due to the general populace not willing to support hosting these facilities due to emission concerns and costs.

In this context, there is a need for policy makers to re-structure their legal and financial framework by capturing the full benefits of effective waste management technologies through a long-term vision that properly recovers the costs from the waste generators.

Landfill fees of $12/ton in Idaho versus $60/ton in New York City create a totally different set of dynamics notwithstanding the differences for example, in the cost of living, economies of scale, access to markets for reuse and recycling and cost of energy. For any community to embrace sustainable solutions, the costs can be significant. Economic justification of any solid waste management method over landfills is essentially impossible on a sustainable basis. Certainly, if the aggregate value of recyclables is over $150/ton or the avoided cost for power production is well over $0.15/kW, the alternative from landfilling appears more favorable. However, this is short-lived since commodity pricing is far more volatile than landfilling. In fact, landfill costs have largely remained steady for decades in all regions until disrupted by regulation that changes the economic advantages of landfills. It is then that communities more aggressively seek sustainable solutions. However, these forces are often not enough. For example, even when regulations 30 years ago required the closing of Fresh Kills Landfill in NYC causing disposal costs to exceed $100/ton, the disruption was short-lived. Rather than seeing costs continuing to climb from $100/ton, New Yorkers saw costs drop to about $60/ton or even less within ten years and today the market cost is still in the $60/ton range for transportation and disposal of waste in NYC. It seems the market eventually adjusts to the regulatory disruption revealing that regulations alone are insufficient.

Public behavior and division of responsibilities

A major outcome of the discussion in all the workshop sessions of this conference was the conclusion that human behavior plays a key role in paving the path towards a more sustainable future for waste management. In line with the concept of Circular Economy as put forth by the Ellen MacArthur Foundation, it is the responsibility of industry to manufacture products that take into account consumer behavior and the end-of-life of the product. In addition to edification of the industry on its role in waste sustainability, the general public has to take part in this education as well. The public must responsibly sort, separate, and dispose of their waste to enable waste infrastructure to effectively work. In recent years, manufacturers have engineered products that are lighter in weight and are far more recyclable. For example, ink printing directly on plastic containers has been almost exclusively replaced by using labels that can be removed. Another example has been the elimination of multi-material packaging such as the earlier soda bottles that used an HDPE base cup adhered to a PET body with an aluminum screw cap. Currently the entire bottle is PET and caps are either HDPE or PP, both of which can be easily sorted during the recycling process.

Currently, there are incentives given to consumers to reduce or reuse in some market areas such as "pay-as-you-throw" programs, however these initiatives have made a modest inroad. More effective universally are the efforts to motivate packaging producers to re-engineer and re-tool their facilities so as to be environmentally mindful during the production cycle. As an example, today institutional investors are far more selective in their investment choices. If they are not encouraged by the proactive steps of a manufacturer to reduce waste, reduce packaging weight and produce products that are recyclable, these investors are turning away. This type of economic influence, although indirect, has had significant impact to drive down the costs of recycling thereby closing the gap with landfills. This is not always the case, but it is clear this trend is still early and more is being done on the manufacturing side than through government taxes or incentives; this reflects the essence of our government and economic structure. One important finding that came of the workshop breakout sessions was that waste management solutions should be economically or regulatory driven. In general, the majority of the public would not voluntarily contribute to waste management and would only pay more unless there is a legal mandate behind it. Programs should be applied more often in the US, such as charging per plastic bags at the supermarket, as is done in most of the EU countries.

There is a lack of education and interest in waste management from the general population. People tend to ignore what they do not know or oppose what they fear. For example, waste to energy solutions are routinely opposed by overburdened communities overpopulated with industry or where air quality issues previously existed. Emissions transparency could help with the public perception in some of these cases. Also, academia can contribute to change the public`s behavior by education. Community acceptance and engagement is critical to inform the public of choices they have in sustainably managing their waste. However, some of the issues with waste management are complex and it is very difficult to simplify some of these problems to the point where the public quickly understands them.

Another issue to consider when trying to change the public`s perception is the aesthetic appeal of some of these facilities, especially in urban areas. Therefore, architectural and structural engineering input are essential, yet there is a gap between different engineering disciplines, municipalities, and developers involved in the design and build of these facilities. One example is the Spitteleau WTE facility in Vienna which is considered a true work of art and is now part of the city skyline and also offers guided tours.

Waste hierarchy

The waste management hierarchy provides the guidelines for a sustainable waste infrastructure. In the hierarchy, waste practices are listed from top to bottom with the most sustainable solution at the top and the least at the bottom. The waste practices listed from top to bottom on the hierarchy developed by the US EPA are as follows: Reduction & Reuse, Recycling/Composting, Energy Recovery, and Treatment & Disposal. "Recycling/Composting" includes anaerobic and aerobic digestion, "Energy Recovery" includes WtE, gasification and pyrolysis, and "Treatment & Disposal" includes landfilling with and without methane capture.

Solutions at the top of the pyramid (reduce, reuse and recycle) are the responsibility of the consumer and producer, while the bottom of pyramid are responsibilities of the waste industry. However, the gap between consumers, producers and waste handlers precludes an efficient waste management system. This is also due to packaging end life being the responsibility of the consumers while producers do not receive incentives to make recyclable products.

Higher reduction and reuse rates can be achieved with help from NGOs and foodbanks, as well as academia, as previously mentioned. However, the situation is more complicated when it relates to food. Food handling regulations are not straight forward, and they vary depending on the types of food that will be reused even if the end product is animal feed. In addition, laws that would protect the food waste generator from the reused food are difficult to control. Health has a higher priority than greenhouse gas emissions, as society tends to consider immediate danger of contaminated or bad food and not the longer-term impact of food waste disposal on the environment.

Packaging has evolved, while materials are better and stronger, lighter, smaller and more cost effective. However, the main drive behind the changes in packaging are the lower production and transportation costs for the producers and not the end life of the products. There are many examples of stores that focus on selling food in bulk to reduce the number of containers used, mostly in Europe, but in many instances, they might end up throwing away more food that spoils faster. Small amounts of plastic can help increase shelf life and decrease spoiled food disposal, for example the plastic foil on cucumbers or other vegetables that prolong the shelf life of certain foods. Rather than focusing on reuse only, a rough guideline should be put in place explaining the tradeoffs of using some plastics and packaging.

This concept can be expanded on many other waste management issues that require a more careful analysis before setting up goals in the waste hierarchy. For example, before transporting food waste from the generator to a composting facility located far away, tradeoffs between benefits of composting and emissions or fuel consumption for transportation should be considered.

The metrics to set recycling goals are outdated and instead, cost benefit analyses should be conducted. The waste hierarchy pyramid is in many instances broken, sometimes reuse is less preferable than recycling, for example with old and less efficient products. The waste hierarchy was developed based on CO_2 equivalent; however the carbon life analysis of each process is required to develop a more accurate hierarchy. Also, policies developed should consider full life cycle analysis of products. The changes in packaging design should lead to changes in life cycle processing and packaging and designers should be involved in regulation development conversations.

Recycling rates are not always meaningful parameters to monitor, the contamination levels in the recyclables are extremely important as well. High recycling rates could also mean stockpiling of recyclables. The waste hierarchy is not helpful for public perception and does not involve the consumer at all levels. The focus is on the 3Rs (reduce, reuse, recycle), however what happens once the trash leaves their homes or business is often not well understood by the consumer.

Technical challenges

There is a clear advantage of treating the waste where it is generated, by avoiding high transportation cost or associated emissions, which in some cases are much higher than the emissions at WtE stacks. Yet having facilities in the city is difficult to accomplish especially for megacities like New York.

NYC has been improving its waste management by looking at anaerobic digestion (AD) solutions, usually implemented at wastewater treatment plants. AD is a suitable solution for treating and recovering energy from food waste but there is a waste by-product that must be dealt with from this type of process. Source separation at the front end is a major factor in the overall process efficiency of converting food waste into biogas. However, food waste collection and source separation in high density areas face many challenges.

Officials and regulators have to deal with the aforementioned issues in megacities, however when facing more pressing issues, such as homelessness, the priorities are different. Nevertheless, NYC is a relatively affluent city so cost should not be an issue or used as an excuse to avoid improving waste management and implement new programs.

Control on the quality of source separation as well as the amount of contamination in the recycling bin is the responsibility of consumers however, government has the significant responsibility to provide consistent and effective education followed by excellent collection services, audits driving continuous improvement, and enforcement. The City of New York has done an excellent job compared to most other cities and has substantially improved the recyclability of the collected materials as well as created a highly effective contracting relationship with a private party that has some of the highest recovery rates in the US.

ACKNOWLEDGMENTS

The authors gratefully acknowledge all the attendees and participants of the Megacity Waste Management workshop in 2016 at The City College of New York. Their expertise and varied perspective combined with respectful and open discussion has enabled the authors to assemble this document. Importantly, those attending the workshop are dedicated to improving the world around them through their efforts focused on sustainable waste management. Without their continued everyday efforts and their willingness to share their expertise and experience, the workshop could not have been possible and this report, documenting some of the outcomes, could not be possible.

REFERENCES

[1]   D. Hoornweg and P. Bhada, "What a Waste. A Global Review of Solid Waste Management," *Urban Dev. Ser. Knowl. Pap.*, vol. 281, no. 19, p. 44 p., 2012.

[2]   D. of S. of N. Y. City, "Annual Report: New York City Curbside and Containerized Municipal Refuse and Recycling Statistics," 2017.

[3]   R. R. and D. S. Casey Cullen, Eric Fell, "An Integrated Waste-to-Energy Plan for New York City," 2013.

[4]   D. Laner, M. Crest, H. Scharff, J. W. F. Morris, and M. A. Barlaz, "A review of approaches for the long-term management of municipal solid waste landfills," *Waste Manag.*, vol. 32, no. 3, pp. 498–512, 2012.

[5]   D. C. Wilson, "Development drivers for waste management," *Waste Manag. Res.*, vol. 25, no. 3, pp. 198–207, 2007.

[6]   A. Demirbas, "Waste management, waste resource facilities and waste conversion processes," *Energy Convers. Manag.*, vol. 52, no. 2, pp. 1280–1287, 2011.

[7]   U. States and E. P. Agency, "Advancing Sustainable Materials Management: 2014 Fact Sheet," 2016.

[8]   D. K. Sharma, S. Bapat, W. F. Brandes, E. Rice, and M. J. Castaldi, "Technical Feasibility of Zero Waste for Paper and Plastic Wastes," *Waste and Biomass Valorization*, vol. 0, no. 0, pp. 1–9, 2017.

[9]   L. Rigamonti, A. Falbo, and M. Grosso, "Improving integrated waste management at the regional level: The case of Lombardia," *Waste Manag. Res.*, vol. 31, no. 9, pp. 946–953, 2013.

[10]   F. D. of E. Protection, "Annual Report 2012," 2012.

[11]   C. Martin-rios, C. Demen-meier, S. Gössling, and C. Cornuz, "Food waste management innovations in the foodservice industry," *Waste Manag.*, vol. 79, pp. 196–206, 2018.

[12]   G. Ionescu, E. C. Rada, M. Ragazzi, C. Mărculescu, A. Badea, and T. Apostol, "Integrated

municipal solid waste scenario model using advanced pretreatment and waste to energy processes," *Energy Convers. Manag.*, vol. 76, pp. 1083–1092, Dec. 2013.

[13]   M. Haupt, T. Kägi, and S. Hellweg, "Data in Brief Life cycle inventories of waste management processes," *Data Br.*, vol. 19, pp. 1441–1457, 2018.

[14]   P. S. Calabrò and M. Grosso, "Bioplastics and waste management," *Waste Manag.*, vol. 78, no. 2018, pp. 800–801, 2018.

[15]   A. Bernstad Saraiva, R. G. Souza, C. F. Mahler, and R. A. B. Valle, "Consequential lifecycle modelling of solid waste management systems – Reviewing choices and exploring their consequences," *J. Clean. Prod.*, 2018.

[16]   L. Hénault-Ethier, J. P. Martin, and J. Housset, "A dynamic model for organic waste management in Quebec (D-MOWIQ) as a tool to review environmental, societal and economic perspectives of a waste management policy," *Waste Manag.*, vol. 66, pp. 196–209, 2017.

[17]   R. Hoogmartens, J. Eyckmans, and S. Van Passel, "Landfill taxes and Enhanced Waste Management: Combining valuable practices with respect to future waste streams," *Waste Manag.*, vol. 55, no. July 2015, pp. 345–354, 2016.

EMERGING APPROACHES FOR WASTE-TO-ENERGY: POTENTIAL OF INNOVATIVE TECHNOLOGIES AND SYNERGIES

Ralf Koralewska, MARTIN GmbH für
Umwelt- und Energietechnik

ABSTRACT

Thermal treatment of waste using grate-based systems has gained world-wide acceptance as the preferred method for sustainable management of residual waste. In order to maintain this position and respond to new challenges and/or priorities, it is necessary to develop and implement innovative technologies. In addition, synergies between different waste treatment processes and/or technologies and various waste streams must be determined.

An optimized multi-stage combustion process has been developed and implemented to significantly reduce the NOx values downstream of the combustion system via primary measures. Additionally, the Shock Pulse Generators and the MARTIN Online Cleaning System are successful fully automatic systems to clean the boiler radiation passes during operation.

Traditional data-processing application software is inadequate to deal with the complex data sets of Waste-to-Energy plants. Therefore, Big Data integration platforms provide a concise overview of data, extract value from data and review key parameters.

The combustion residues bottom ash and fly ash contain substantial amounts of raw materials (metals / minerals / phosphorus), which are finite resources. New technologies and processes for the recovery of these reusable materials on-site/off-site of Waste-to-Energy plants had been developed. Sewage sludge mono-combustion plants enable to use the sewage sludge as a source of phosphorus.

Separately collected organic waste requires different treatment processes. MARTIN dry digestion holds the potential for producing high-quality products such as biogas, compost and liquid fertilizer.

Implementing Waste-to-Energy and sewage sludge mono-combustion or dry digestion in the same location offers many potential possibilities for synergies. Various material and energy flows can be united in a meaningful way, which in turn results in additional benefits for these types of plants.

This paper describes the main progresses achieved and innovative technologies implemented in large-scale by MARTIN. Also, in the future all types of waste will be treated in accordance with ecological and economic constraints and in compliance with legal requirements.

INTRODUCTION

Waste treatment technologies have been confronted with numerous, varying challenges over the last decades. Any new demand leads to an increase in the overall complexity of the treatment processes. Additional skills and competence are needed for plant design, process control and operation. Innovative technologies and synergies between different waste treatment processes and/or technologies successfully provide solutions for optimizing in terms of climate and resource protection, reduction of environmental impacts as well as political, regulatory and market aspects.

LOW NO$_X$ TECHNOLOGIES

Most of the waste's nitrogen content is transferred to the flue gas during combustion as nitrogen oxide NO$_x$. The limit values for NO$_x$ emissions continue to decrease as a result of regulatory requirements. Compliance with limit values is possible with the SNCR (Selective Non-Catalytic Reduction) process, which injects ammonia or urea into the furnace. In some cases, SCR (Selective Catalytic Reduction) catalytic converters are necessary. However, these involve higher costs (investment / operation) and energy consumptions.

In this respect, various concepts of low NO$_x$ technologies were developed to significantly reduce the NO$_x$ values downstream of the combustion system via primary measures. This is due to the fact that chemical reactions that convert the primarily formed NO$_x$ back to nitrogen are promoted as a result of the reduced excess air and consequently higher temperatures in the lower area of the furnace.

For a conventional combustion system setting the underfire air is set to be slightly super-stoichiometric. An excess air rate is achieved by supplying overfire air for flue gas burnout. The innovative concept deals in particular with the positioning/distribution of the overfire air nozzles and options for mixing the uncombusted gases in the post-combustion zone as a measure for NO$_x$ reduction combined with the SNCR process, which injects ammonia or urea into the furnace. With the modified overfire air configuration in the upper furnace area the NO$_x$ content can be reduced significantly. It has been proven under continuous commercial conditions that NO$_x$ values are reduced to less than 80 mg/Nm^3 (11 % O$_2$, dry) by primary measures plus injection of ammonia or urea. A further feature of this process is that a NH$_3$ slip of less than 10 mg/Nm3 is reached at the same time.

Nevertheless, super-stoichiometric conditions are reached at the overfire air level. The 850°C / 2 seconds - criteria for flue gas residence time, based on EU legislation, is fulfilled starting from the last injection level in the furnace. The reduced excess air rate allows cost reduction in the boiler and flue gas cleaning and improved boiler efficiency. A further advantage is the reduced flue gas velocity in the lower furnace which leads to a reduction in the fly ash carried over to the boiler.

ONLINE BOILER CLEANING

In Waste-to-Energy plants, efficient online boiler cleaning is of major importance because severe fouling or even clogging of heating surfaces can otherwise occur after short operating periods. The deposits build up on the boiler heat exchanger surface results from mostly inorganic gaseous, liquid and solid products of the combustion process.

The deposits lead to a number of negative impacts on the Waste-to-Energy plant operation:
- Reduction of the heat transfer rate / energy efficiency
- Increase of the pressure drop along the flue gas path
- Reduction of operating period
- Potential increase of corrosion

The MARTIN Online Cleaning (MOC) System is a fully automatic system to clean the boiler radiation passes (1st, 2nd, 3rd boiler pass) during operation using water to keep the temperature at which the flue gas enters the convective heating surfaces as low as possible and thus achieve longer service periods (Fig. 1a).

The Shock Pulse Generator (SPG) is an online boiler cleaning equipment automatically generating shock pulses by pressurized, controlled gas combustion (Fig. 1b). The supersonic shock pulse is discharged into the boiler, puts the gas volume and boiler surfaces into a short oscillation, causes an impact sound wave within the deposits and therefore cleans off the fouling in a very effective way. The frequency of the online boiler cleaning depends on the fouling behaviour of the boiler system and the boiler system operation.

Both technologies have proven its outstanding performance at worldwide different installations. Plant operators confirmed significant extension of boiler operating periods and increased efficiency, thus contributing to a sustainable and economic plant operation.

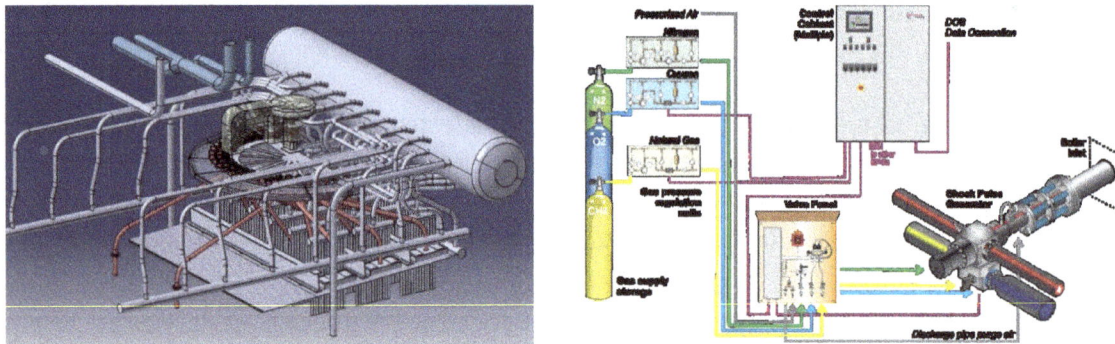

Fig. 1: (a) MARTIN Online Cleaning (MOC), left, and
(b) Shock Pulse Generator (SPG), right

BIG DATA

Today, high plant availability rates and low emission levels are the state of the art in the Waste-to-Energy industry. To make further progress in these fields, it does not suffice to simply display and record plant data. It is rather necessary to intelligently evaluate the individual plant conditions.

Waste-to-Energy plants capture and record all process and emission-relevant data. The resulting mass of data is too large to efficiently evaluate it manually. One year of plant data recorded in a Waste-to-Energy plant, for example, correspond to approx. 40 GB of data. Using suitable algorithms, this abundance of data ("big data") can be used to retrieve meaningful additional information for the plant ("smart data").

Data sets of Waste-to-Energy plants are so big and complex that traditional data-processing application software is inadequate to deal with them. Big data includes capturing data, data storage, data analysis etc. Integration platforms provide a concise overview of data, extract value from data and review key parameters to consider the optimization of supply chains and predictive maintenance options.

On the one hand, such assistive systems can be used immediately during plant operation for controlling or cleaning the boiler (by forecasting fouling). On the other hand, the systems can be used for simplifying maintenance and spares coverage for the plant ("predictive maintenance"). Therefore, the stable operation of a Waste-to-Energy plant can be guaranteed in addition to periodic maintenance or repair of equipment by after sales service.

By now, analysis tools for large amounts of test and operating data are also essential for developing plants further and are the basis to render possible the erection of efficient Waste-to-Energy plants with low emission levels and high availability rates.

Online balancing is the first step towards purposefully recording and evaluating operating data and involves computing material, mass and energy balances for the entire plant and its individual process components. Computing the balances indicates the plant's current condition and establishes the potential for optimizing the plant's efficiency and capacity.

RECOVERY OF RECYCLABLES

Sustainable processes for the recovery of recyclables are becoming increasingly important in view of the continual depletion of raw materials. The combustion residues from grate-based Waste-to-Energy plants / mono-combustion plants (fluidized bed) contain substantial amounts of raw materials (metals / minerals / phosphorus), which are finite resources and whose primary production results in high power consumption and high environmental impacts.

In conventional grate-based WTE plants, bottom ash is removed from the furnace via a wet-type discharger. Dry discharge of bottom ash can increase the recovery rate of ferrous and non-ferrous metals significantly. MARTIN developed a system of dry bottom ash discharge (Fig. 2) which consists of the following components:

- MARTIN ram-type discharger
- Air separator enclosed in container
- Dust extraction system
- Air system

The discharger is operated without water. The dry-discharged bottom ash is conveyed directly to an air separator. The fine fraction and bottom ash dust is extracted in a defined manner. The air separation area is enclosed by a container preventing false air from entering the furnace and dust from getting into the boiler house. The exhaust air from the air separator is conveyed to a dust extraction system, where the fine fraction is separated. The unburdened air could be conveyed to the combustion air system.

Various advanced techniques and concepts for the recovery of resources out of bottom ash have been developed and installed in large-scale implementations on-site Waste-to-Energy plants and processing facilities. Typically, ferrous and non-ferrous bulk metals are recovered but also precious metals, glass and stainless steel are of great interest. The reuse of the mineral fractions varies in different countries.

The MARTIN bottom ash treatment system (MARTIN Slagline; Fig. 3) suitable for wet- and dry-discharged bottom ash is modular in design and comprises the following components:

- Tipping hall / material feeding system
- Grading

- Removal / refining of ferrous metals
- Removal of non-ferrous metals / stainless steel / glass (optional)
- Receiver tanks for mineral residues

The process of selective zinc recovery from the acid-scrubbed fly ash from Waste-to-Energy plants (Fig. 4) is one example of a process-integrated method for recovering economically profitable heavy metals. Cadmium, lead and copper are separated using a reductive process and recovered as a metal mixture in lead works. Zinc is separated from the pre-cleaned filtrate using a selective extraction method, and then concentrated and recovered electrolytically as pure zinc (Zn > 99.995 %).

The synergies associated with the residues occurring with wet flue gas cleaning are used during the process. During acidic ash extraction, the heavy metals in the fly ash are mobilized and extracted by the acidity of the quench water. At the same time, the excess acid content of the quench water is neutralized by the alkalinity of the fly ash. After acidic fly ash scrubbing, the filter ash cake has an extremely low heavy metal content. Any organic matter that remains in the cake subsequent to scrubbing can be returned to the combustion system so that it can be destroyed.

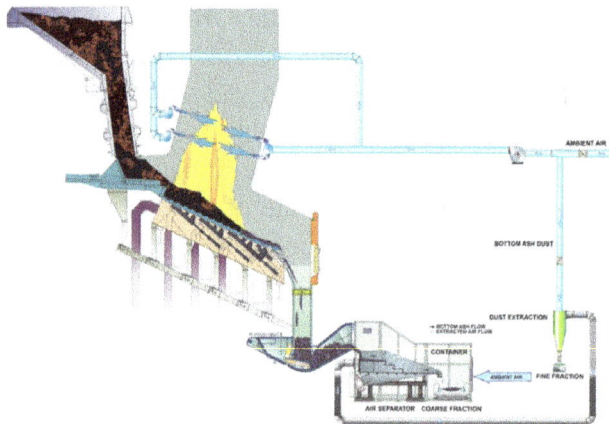

Fig. 2: MARTIN dry bottom ash discharge

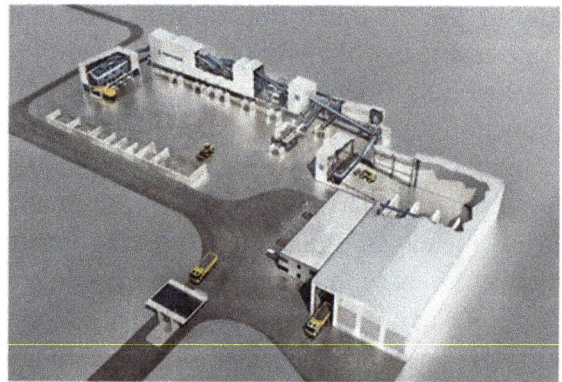

Fig. 3: MARTIN Slagline (bottom ash treatment facility)

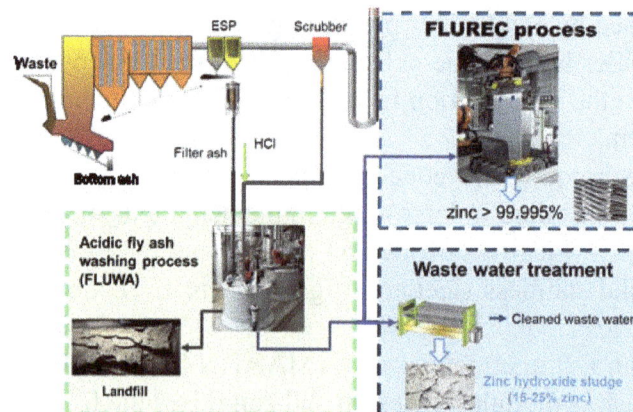

Fig. 4: Fly ash treatment / FLUREC process

In Switzerland filter ashes are treated by acidic ash extraction transferring the metals in a zinc-containing hydroxide sludge. The rising costs for the disposal and limitations of treatment in zinc smelteries (costs / acceptance by smeltery operators) are making the zinc recovery from filter ashes increasingly difficult. The SwissZinc project tries to overcome these obstacles and proves the technical and economic feasibility of direct materials recovery. Presently, the development and establishment of a central large-scale processing facility for a metal recovery from hydroxide sludge is going on.

Sewage sludge mono-combustion plants enable to use the sewage sludge as a source of phosphorus. The KÜTTNER MARTIN Technology GmbH offers the fluidized bed technology for the thermal treatment of sewage sludge. Technically pure, heavy-metal-free, commercial phosphoric acid or other fertilizer raw material can be produced. A sufficient supply with steam and/or energy from a Waste-to-Energy plant nearby supports synergy effects.

DRY DIGESTION

Due to their composition, residual and organic waste require different processes when it comes to their treatment. In particular separately collected organic waste holds the great potential for producing biogas, compost and liquid fertilizer. In view of this, MARTIN offers dry digestion plants (Fig. 5). The biogas can be used to generate electricity and/or heat in combined heat and power plants or it can be supplied to the natural gas grid following treatment. Dewatering the fermentation residues produces high-grade compost as well as valuable liquid fertilizers.

The core component of dry digestion plants is the digester, in which the organic waste is treated under anaerobic conditions. The continuous thermophilic process ensures complete conversion of organic waste into biogas and safe hygienization of the fermentation residues. For this conversion process, MARTIN uses THÖNI's plug flow digester on an exclusive basis. The process inside the digester is not negatively affected by high amounts of unwanted materials so that both organic and green waste as well as the organic fraction from residual waste can be treated.

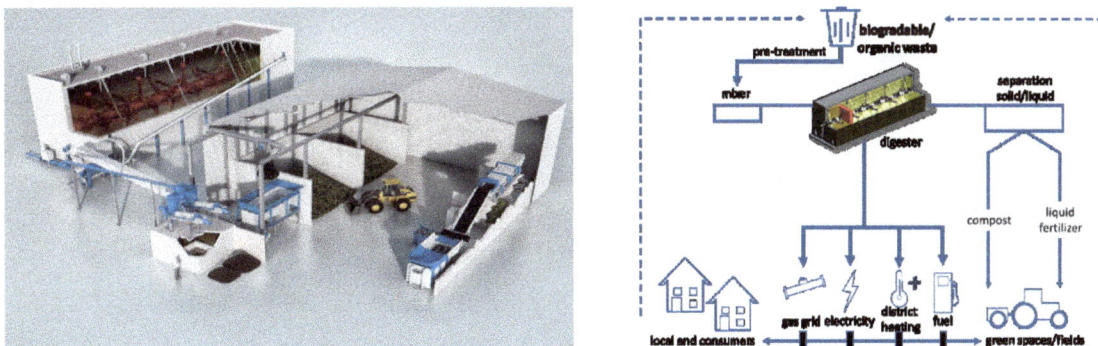

Fig. 5: MARTIN dry digestion (Thöni system)

Implementing Waste-to-Energy plants and dry digestion plants in the same location offers many possibilities for synergies. Various material and energy flows can be united in a meaningful way, which in turn results in additional ecological and economic benefits for both types of plants. Necessary infrastructure such as weighing devices, road systems, supply facilities, disposal systems and staff may be used for the benefit of both plant types. The Waste-to-Energy plant ensure that digestion plants are reliably supplied with the electricity and heat required for daily operation. One plant, where these potentials are already used, is located in Augsburg (DE). The untreated biogas is processed and supplied to the natural gas grid as biomethane.

CONCLUSIONS

Thermal treatment of waste using grate-based systems has gained world-wide acceptance as the preferred method for sustainable management of residual waste. But in order to maintain this position and respond to new challenges and/or priorities, it is necessary to develop and implement innovative technologies. In addition, synergies between different waste treatment processes and/or technologies successfully provide solutions for optimizing in terms of climate and resource protection, reduction of environmental impacts as well as political, regulatory and market aspects. Combining Waste-to-Energy plants and other treatment facilities (dry digestion / sewage sludge mono-combustion) in one location facilitates the implementation of modern residue recovery centres, which not only ensure up-to-date treatment but also focus on an ideal recovery of recyclables.

Further development and optimization of existing technologies and concepts are needed due to international requirements. It has been repeatedly proven that innovative technologies must first be developed and comprehensively investigated. Additional skills and competence are needed for plant design, process control and operation. Nevertheless, cost-benefit and eco-efficiency analyses clearly show that these additional efforts should be made. In the future, MARTIN will continue to reliably ensure treatment of waste under ecological and economic constraints, using innovations and reliable process engineering technology and taking international statutory requirements into account.

In selected countries, including the USA, MARTIN® is a registered trademark. MARTIN technologies are covered by numerous patents.

A CASE STUDY OF GASIFICATION CHP IN NORTHERN ITALY IN THE EUROPEAN CONTEXT AND COMPARISON TO TRADITIONAL COMBUSTION SYSTEMS

Marco Baratieri, Free University of Bolzano

INTRODUCTION

There are several technologies that recover energy from waste, but the term "waste-to-energy" specifically describes thermal treatment of municipal solid waste (MSW). The most common thermal treatment technology is incineration [1]. In Italy, approximately 6 million tons of municipal solid waste are incinerated annually at waste-to-energy plants. This value has increased by almost 40 % in the last decade, despite fewer plants in operation. As seen in Fig. 1 the total number of waste-to-energy plants has decreased from 49 to 44 during the last decade but this has not affected the treatment capacity as the total amount of waste has increased.

The development of larger capacity waste-to-energy plants is strategic for increasing the overall electricity production for this sector since bigger scale facilities use larger steam turbines with higher isentropic efficiencies. Additionally, the implementation of larger facilities allows for the optimization of the waste transportation schemes and the overall supply chain, which can assist the management of operational costs. A peculiar case is the waste-to-energy plant in Brescia, which in 2006 was accredited by Global Waste-to-Energy Research and Technology Council (WTERT) as the best waste-to-energy plant in the world. It has three energy production lines and a net electrical efficiency of 27 %. This plant also impressively covers approximately 75% of the city's heat demand.

Aside from higher efficiency demand, waste-to-energy plants have assisted the aversion of environmental crises such as the one which occurred in Naples. The Naples waste management crisis can be roughly described as "a series of events" caused by the lack of waste collection in the Campania region between the years 1994 to 2012. As a result, hazardous waste chemicals and heavy metals were dumped on public roads and in squares increasing the risk for incidents that could potentially compromise public health. The development of a waste-to-energy plant in nearby Acerra, along with other supporting measures, helped gradually de-escalate the situation and radically improve environmental conditions. The local waste-to-energy plant has been operating at full capacity since September 2009 and can incinerate approximately 2,000 tons of municipal solid waste per day. The plant has a nominal electrical production of 120 MWe. In addition, it has incorporated the best available APC (i.e. air pollution control) technologies in order to mitigate any potential dangers.

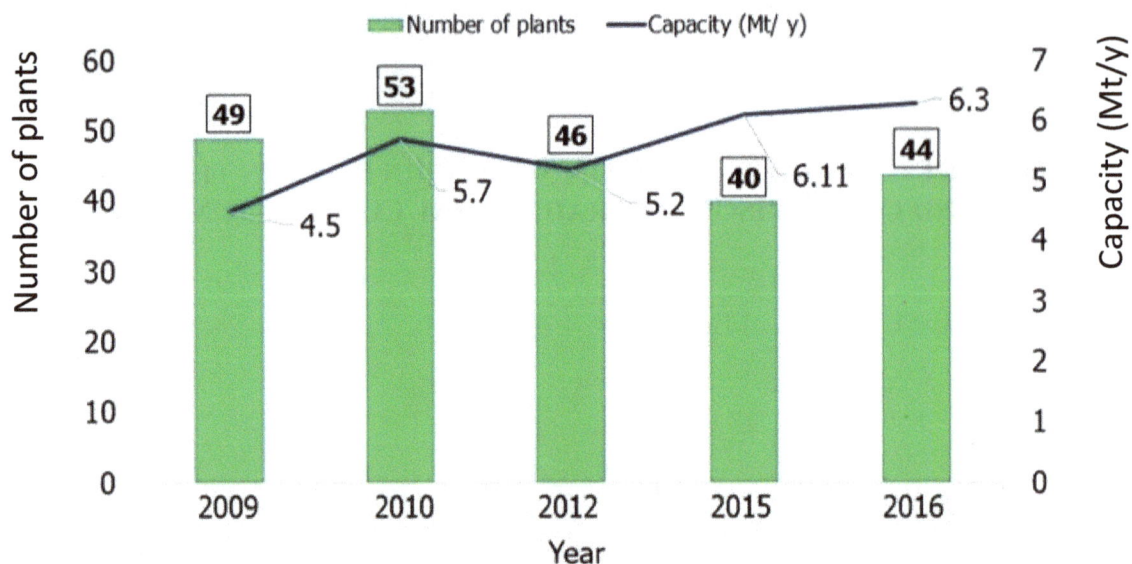

Fig. 1: Waste-to-energy plants operating in Italy and their capacity

LIMITATIONS OF INCINERATION AND THE CASE FOR GASIFICATION

While incineration of MSW has several advantages, it also has a few limitations. In particular, the size of the plant is crucial for maximum electrical efficiency. The management of bottom and fly ash, a byproduct of solid waste combustion, remains a challenge. Also the utilization of waste heat is not always straightforward, especially in Mediterranean countries which have a relatively hot climate. During the years of 2004–2007, CHP (i.e. combined heat and power) facilities sometimes were not even connected to district heating networks.

Gasification can be an interesting solution to be investigated due to the increased electrical efficiencies – especially at the small scale – and due to the limited environmental impacts regarding gaseous emission with respect to traditional combustion. For the case of biomass utilization, Italy and other Central-European countries have installed several gasification facilities. Here, the example of a facility in South Tyrol, Italy can be used as a case study.

As shown in Fig. 2, approximately 50 small-scale biomass gasification plants – i.e. 50 to 400 kW electrical power – have been developed in the region of South Tyrol since 2012. Gasification, contrary to combustion, is a thermal process that uses sub-stoichiometric air to convert a carbon-rich fuel into (mainly) gaseous fuels by thermo-chemical mechanisms. Several different gasification technologies are available and are being developed. Most of the current plants utilize fixed-bed reactors, treat wood chips, and are coupled with Otto engines for CHP production. The reasoning behind these choices are that the reactors are easy to operate and produce syngas (or producer gas) with low tar content. Additionally, the use of Otto engines is correlated to lower nitrogen oxides, NOx, emissions and which allows operators to manage their CHP engines solely with the gasification exhaust gas without the necessity for burning as a secondary fuel. This, for example, would be the case for the operation of a diesel engine coupled with the gasifier.

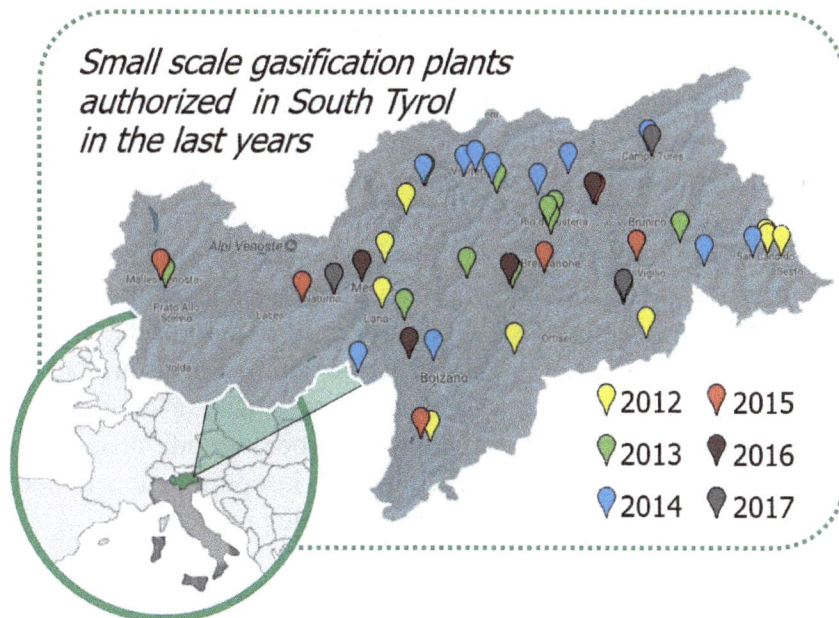

Fig. 2: Development of newly authorized small-scale gasifiers in South Tyrol, Italy

PLANT MONITORING METHODOLOGY

The measurement of these small-scale gasification units has been a challenge and only few studies can be found in the literature on this topic. Patuzzi et al. [2] implemented a thorough monitoring campaign and used the "CTI 13 Recommendation" as a standard reference for the monitoring activities. The Italian Committee of Thermal Engineering issued this recommendation, which consists of guidelines for contracting and commissioning of gasification systems that produce and utilize producer gases by gasification of biomass of lignocellulosic structure. Along with "Recommendation CTI 13", the authors also used other European standards like the standard EN 14778:2011 for biomass sampling and the technical specification CEN/TS 15439, commonly known as "tar protocol". The quality of the gas was measured onsite with a mobile gas chromatography unit and the mass balances were measured on an input-output basis. These parameters assisted the implementation of energy balances for all the representative plants. Vakalis et al. [3] presented the whole design of the monitoring campaign.

RESULTS AND DISCUSSION: QUALITY OF THE PRODUCER GAS
AND POTENTIAL FUTURE DEVELOPMENTS

The quality of the producer gas from the above mentioned small-scale gasifiers has significantly improved in comparison to older gasifiers. Fig. 3 shows the range of concentrations of the major gases comprised in the producer gas and the variability of the lower heating values – obtained as a weighted average – for all measured small-scale gasification plants in South Tyrol. The improved designs allowed these gasifiers to operate with less amounts of air and this is reflected in the concentration ranges for nitrogen. These gasifiers usually operate in sub-atmospheric conditions which favor higher concentrations of hydrogen and methane. The concentration of hydrogen is also liberated during staged drying, a pretreatment step prior to feeding into the gasifier, designed to dehumidify the biomass. Additionally, the optimization of automation and controls has assisted operators to achieve higher temperatures in the combustion zone than previously. This has increased the concentrations of carbon monoxide, which is produced not only from the incomplete combustion but also secondarily from reactions between carbon dioxide and char. Considering these measurements, the lower heating values (LHV) of producer gases from modern small-scale gasifiers is approximately 5 ± 0.5 MJ/kg.

A final point to be made about the operation and management of small-scale gasifiers is related to the management of the residual char. This is the solid by-product of gasification and for the small-scale biomass gasifiers consists of 2-5 weight % of the initial input biomass. Char valorization is currently of great interest. One conventional approach is the use of char in agricultural fields as soil enhancer. Several novel utilization scenarios are being investigated at the moment, such as the use of char as a catalyst for tar cracking or for gas reforming. The characteristics of the produced char is influenced by the process conditions, which, nowadays are designed to maximize the energy generation. Multi-criteria optimization is needed in order to increase the quality of the residual char and calibrate suitable productions.

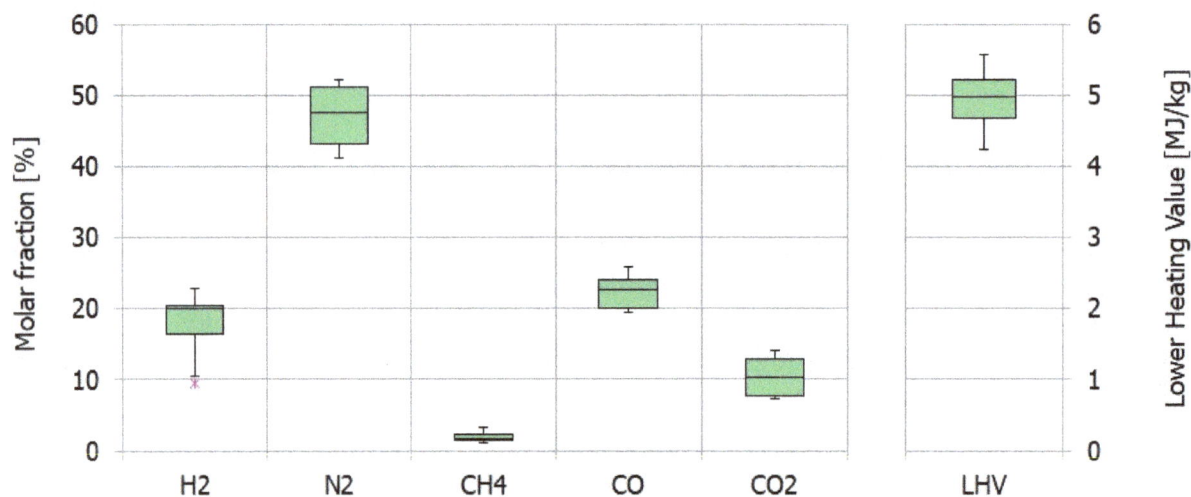

Fig. 3: Range of concentrations for the major gases in the producer gas of small-scale gasifiers in South Tyrol

REFERENCES

[1] S. Vakalis, K. Moustakas and M. Loizidou (2018). Assessing the 3T method as a replacement to R1 formula for measuring the efficiency of waste-to-energy plants. Waste Management & Research 36, 810 – 817.

[2] F. Patuzzi, D. Prando, S. Vakalis, A.M. Rizzo, D. Chiaramonti, W. Tirler, T. Mimmo, A. Gasparella, M. Baratieri (2016). Small-scale biomass gasification CHP systems: Comparative performance assessment and monitoring experiences in South Tyrol (Italy). Energy 112; 285-293.

[3] S. Vakalis, D. Prando, F. Patuzzi, T. Mimmo, A. Gasparella, W. Tirler, K. Mair, G. Voto, D. Chiaramonti, A. Rizzo, M. Pettorali, M. Prussi, S. Dal Savio, D. Andreasi, M. Baratieri. Assessment of a test methodology suitable for small scale biomass gasifiers. 22nd European Biomass Conference and Exhibition. 23rd - 26th June 2014, Hamburg.

UTILIZATION OF RESIDUAL LIGNOCELLULOSIC RESIDUES AND CO₂ IN CANADA: THE BEGINNING OF A WHOLE NEW ERA OF OPPORTUNITY

Michael Lugo-Pimentel, Université de Sherbrooke, Québec

Xavier Duret, Université de Sherbrooke, Québec

Bruna Rego de Vasconcelos, Université de Sherbrooke, Québec

Thierry Ghislain, Université de Sherbrooke, Québec

Jean-Michel Lavoie, Université de Sherbrooke, Québec

ABSTRACT

Canada has been relying on the integration of its forest industry for decades during which lumbering was closely associated with the pulp and paper industry back in an age where CO_2 emissions were merely discussed. Nowadays, the country is facing many challenges, one of which being the progressive disappearance of the pulp and paper industry, which used to be a pillar of the economy. In addition, Canada has one of the largest levels of greenhouse gases (GHG) emission per capita in the world which can be related to the weather, the area of the country and of course, its sparse population. Despite these issues, numerous opportunities could be considered to reduce the amount of wastes while at the same time contributing in reducing the GHG emissions. In addition, both Provincial and Federal Governments have established different programs contributing in supporting R&D activities both at laboratory scale up to industrial scale. In this short review, the possibility to put together a "less-than-zero" GHG emissions biorefinery using solely wastes and green electricity will be discussed.

INTRODUCTION

With the significant increase in virtual medias across the world, the classical pulp and paper industry (producing newspapers) has been declining in many countries including in Canada [1]. Nevertheless, the lumbering industry is thriving, and large volumes of wood are currently being produced [1]. In Quebec, for example, the average yearly production of wood residues is roughly 6.2×10^6 dry metric tonnes and about 1.9×10^6 tonnes for bark in 2017 [2]. Part of this production is being used as a feedstock for other industries such as the wood composite or the expanding pellet industry. Despite of these uses, the market does not use as much residual biomass as what is available leading to a significant drop in the price of the feedstock, around 2.90-32.50 CAN$ per tonne for wood sawdust and 0-22.5 CAN$ per tonne for bark [3]. The latter has represented and still represents a challenge for the industry since this residual biomass can hardly be used in combined heat and power (CHP) units (high ash content) and the usually high level of extractives makes it quite difficult to be used as a feedstock for a potential biorefinery [4]. Despite the lack of large-scale projects to use such residue in Quebec and Canada at this point, there are still ongoing projects that focus on the utilization of the secondary metabolites found in softwood bark [5], [6]. In the latter case, it could be a very good value-adding approach for bark since it would also offer the opportunity to use the structural part of the biomass for other applications (such as for the energy production).

In addition to the very large volumes of residual lignocellulosic biomass available in the country, Canada also has large quantities of green electricity. The electricity situation varies from one side of the country to the other, but on the east coast, the province of Quebec is currently one of the locations in North America where electricity is among the cheapest [7]. With a current industrial price of 0.0327 CAN$/kWh and consumer price of 0.0591 CAN$/kWh, energy is very if not too available [8], [9]. This abundancy of green electricity is likely to increase in the upcoming years because of the growing and cheap availability of other renewable systems such as photovoltaics and wind turbines [10]–[13]. In addition, it is envisioned that the new generation of homes will be increasingly self-sufficient in energy which could only lead to an increase of the available green electricity [14]–[16]. On the latter, there is a growing concern around the world on how to store this electricity and in addition to the traditional approaches such as batteries and uphill water pumping, different new options are being considered including the Power-To-X (PtX) concept [17]. The PtX concept relies on the utilization of green electricity during off-peak periods in order to produce chemicals such as hydrogen, methane, methanol and liquid transportation fuels such as gasoline, diesel and even jet fuel [17]–[19].

While there is significant surplus of electricity and lignocellulosic biomass, the case of GHG emissions is not entirely solved. Amongst the largest producer of GHG, the oil and gas sectors represent 26% followed by transportation and heavy industry with 25% and 11% of GHG emissions in the country [20]. This situation is quite homogeneous all over Canada, although some slight differences may occur. For example, in Quebec, transportation represents the larger source of GHG while industry comes second [21]. This could be, in part, related to the abundancy of green electricity available in the province. To take even more advantage of the availability of this renewable energy, a strong push has been made towards implementing electric cars. However, in Quebec as in the rest of the country, the population is sparse which implies that such a solution would have an impact in large cities and not on the entire territory, where the travel distances often surpass the autonomy of current EVs. Still, the province has committed towards reducing significantly its GHG emissions (37.5% below 1990 levels by 2030) and in order to do so, drastic measures are required.

In this manuscript, different scenarios will be discussed involving green carbon, renewable electricity and the necessity to reduce GHG emissions which could be combined into a unique opportunity in a model that would respect the concept of sustainable development.

DISCUSSION

Lignocellulosic biomass is typically composed of cellulose, lignin, hemicellulose and secondary metabolites often referred to as "extractives". Both cellulose and hemicelluloses are composed of carbohydrates and while cellulose is essentially composed of glucose units, the hemicelluloses composition will vary from one type of biomass to another and involves different concentrations of C5 and C6 sugars. Lignocellulosic based carbohydrates have been considered for decades as a source of renewable sugars that could be converted to different products, including but not limited to ethanol. While ethanol is the obvious choice in a market that consumes large amount of it on a yearly basis ($2x10^9$ L are mandated in gasoline in Canada yearly), second-generation sugars could represent a significant opportunity for other commodities and fine chemicals, especially the one produced biologically using C6 sugars as feedstock. Compounds such as citric acid, malonic acid, 5-hydroxymethyl furfural, levulinic acid, lactic acid, etc. could also be produced from the same substrate hence increasing the potential outputs that could be generated from lignocellulosic biomass.

One of the classical approaches for producing second-generation ethanol consists of isolating the cellulose macromolecule from the rest of the biomass; an approach that would, to a certain extent be aligned with classic pulp and paper processes. In addition to the classical pulping techniques such as the kraft or the soda pulping, numerous other alternatives have been considered so far in literature in terms of pretreatment in order to isolate the carbohydrate rich substrates. Techniques such as steam explosion [22] and organosolv [23] processes are often reported in literature and are as well considered by many groups as a first step to produce cellulosic ethanol. Independently of the technique (or combination of techniques) used to produce cellulosic ethanol, the reality of the commodity market requires the approach to be as simple as possible to allow the production of cheap second-generation sugars. Lignin also represents a major concern in this industry since it is very abundant in biomass and requires to be used for value-added purpose in order to increase the output value of the biomass being used [24], [25]. Hence, for a process relying on the isolation of cellulose for the production of cellulosic ethanol, it is of utmost importance to reduce to a maximum the amount of unit operations involved. Therefore, a technology that would require only very small processing would be completely beneficial for the economics of the system. Which is why, technologies such as the one being scaled by ReSolve Energy Inc. could represent a real game changer in this industry, allowing the direct conversion of any type of feedstock to fermentable sugars and third generation pellets. This approach is especially interesting because it can even process bark, a very cheap and abundant feedstock in Canada and in Quebec. Because of its ash content and high extractive content, this feedstock does not have numerous opportunities for utilization in the industry and represents a significant challenge for the industry.

The forest industry can be one of many key components in the efforts to reduce GHG emissions as established by the different government mandates. The forest industry is at this point producing large volumes of residual biomass which includes wood chips, sawdust and shavings from the lumbering industry. The volumes have been estimated roughly to $6.2x10^6$ tonnes (dry basis) per year in 2017 [2]. While there is still some uptake from the pulp and paper, the wood panel and the pellet industry, the volumes involved at this point are significant which has led to a general decrease of the value of this biomass down to about 32.50CAN\$/tonne [3]. Bark is also a very abundant biomass with an average production of $1.9x10^6$ tonnes in 2017 with a very low industrial uptake [2]. Hence, the cost of such type of feedstock is currently in the range of 2.90-32.50CAN\$/tonne (dry basis) and is often considered as a waste. On average, the annual consumption of transportation fuel in the province of Quebec is in the range of $11x10^9$ L per year (combining $8.4x10^9$ L of gasoline and $2.7x10^9$ L of diesel – 2016) [26], with 10 vol% of dry ethanol added [27]. If the inventoried forest products were to be used for the production of second-generation ethanol for example, it could provide up to $2.45x10^9$ L of renewable second-generation fuels and using processes such as the ReSolve technology, would also bring $2\text{-}3\mathrm{x}10^6$ tonnes of third generation pellets (26GJ/tonne) to the market. On the transportation fuel side, which is the most important contributor to GHG in Quebec, using the forest residues to this purpose would reduce GHG emissions by roughly 25% below 1990's levels, which has been targeted by the province. Of course, reaching such levels would be very hard even in the case of the full extent of the utilization of such feedstock by the fuel industry (which would not be realistic), it still represents an avenue where different options are required to meet the GHG reduction expectations. Furthermore, agricultural residues could also be added to the equation, reducing the pressure on other industries relying on wood residues and providing alternatives for wastes such as corn stovers.

The distribution of green electricity and the transportation of liquid fuels from in-situ biorefineries represent a challenge in large countries. The cost of green electricity has declined around the globe because in part of large initiatives such as the Energiewende in Germany that can be directly related to a significant decrease of the cost of photovoltaic panels [28]. The energy autonomy of houses is increasing, and it is envisioned that in a nearby future, communities will be able to cope with their own energy requirements as well as adding some more energy to the grid [29]–[31]. In all cases, it is envisioned that the amount of renewable electricity on the market will steadily grow in the upcoming years [32]. However, this new reality brings storage issues leading to a growing necessity in storage opportunities. Among the many options that could be considered, such as pumping water uphill during off-peak periods [33], [34], the chemical conversion remains a very valuable one. The fundamental approach for this remains chemical batteries which have been improved over the years and have shown flexibility of form factors to fit many uses (such as in cell phones or in cars). Consequently, the development of large capacity batteries dedicated to store green electricity is an active area of research [35]. Another option could be to convert the renewable electricity to commodity chemicals. The option of producing hydrogen out of electricity has been investigated for decades but in the past 10 years or so, another concept has been rising in the scientific and industrial community which is the Power-To-X [36], [37]. In the latter, the X represents a large variety of compounds that could be derived from green power. Initiatives such as the Audi power-to-gas (methane) [38] or the Enbridge power-to-gas (hydrogen) [39] are examples of this interest. In addition, other companies such as Carbon Recycling International [40] and Swiss Liquid Future [41] are looking to go from green electricity to liquid fuels such as methanol. With methanol being a known commodity, that could be used directly as oxygenate in engines or its dehydrated form (dimethyl ether) in diesel engines [42]. Reports from the open literature also shows that methanol could be used as building block to produce jet fuel, a possibility that is getting some traction in the scientific community because of the fewer options for this specific type of fuel [43]. While production of hydrogen could be made from the electrolysis of water directly, the production of methane or methanol requires a carbon substrate hence creating a very interesting although challenging pathway towards CO_2 utilization.

CO_2 could be considered as the ultimate carbon waste and it is produced from different sources including from the combustion of classical fuel such as coal, oil-derived products and even biomass [44]. Is it also produced in parallel with most of the technologies aimed at producing carbon-based biofuels such as fermentation, gasification, pyrolysis, torrefaction, HTL, etc [44], [45]. While it may not be directly a side-product of the previously mentioned processes, it can still be a by-product originated from the production of energy required for those processes [46]. The increasing accumulation of carbon dioxide in the atmosphere has been worrying the scientific community for decades and in order to cope for this growing concern, the different nations around the world have committed to control if not reduce their GHG emissions. Reduction of CO_2 generation requires not only deep modifications in the behavior of the global community but will also require technologies that allows reducing carbon dioxide either before or after being generated. In order to stimulate the industrial commitment towards reducing GHG emissions, different countries around the world have committed to create a taxation system. In Canada, for example, starting in 2018 CO_2 was associated to a minimum value of 30 CAN$ per tonne and is expected to rise by 10 CAN$/tonne/y [45], [46]. Other system such as the cap-and-trade system adopted by the State of California and the Province of Quebec provides another type of control on CO_2 as well as creating economic incentives to stimulate its reduction. Overall, providing a negative value to carbon should stimulate its reduction if not its utilization as a feedstock. The combination between carbon dioxide and excess renewable electricity in that sense could represent an ideal solution to combine two problems into a sole opportunity. Carbon dioxide would be sent back directly on the market, hence providing a faster and more efficient utilization of energy as compared to fossil fuel and even biofuels. Fig. 1 below depicts how this can represent an opportunity.

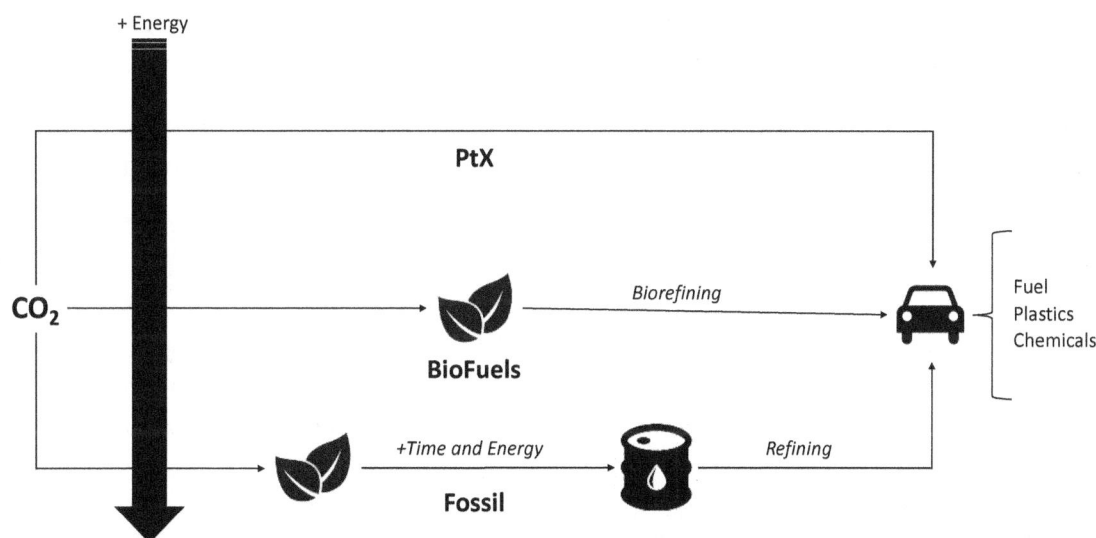

Fig.1: Comparison between an integrated Power-To-X technology combining renewable electricity and carbon dioxide

In light of the potential for combining carbon dioxide and green electricity, the potential of biomass could be not only on the stabilization of the carbon dioxide emissions from transportation fuels but also on the overall reduction of the CO_2 content in the atmosphere. What could be referred to a "less-than-zero" carbon footprint.

For example, referring again to the situation in Quebec (Canada) where the renewable residual biomass is abundant as well as the green electricity, and in a country that has established a carbon taxation mechanism, the utilization of the forest residues could provide a very suitable source of renewable carbon for second generation fuels (assuming the cost-effectiveness of any technology). It is generally considered that second-generation ethanol could lead to an 86% GHG reduction as compared to fossil fuels [49]. If the fuel were to be produced through fermentation, recuperating the CO_2 produced as well during the process (about half of the sugar processed leads to CO_2) for PtX technology could allow reducing the GHG emissions not only to the point that it would be completely carbon neutral, but would as well reduce the overall carbon footprint. Such situation could be considered in Quebec but could as well be considered in countries that allies these three important factors, which are availability of renewable carbon, green electricity as well as carbon reduction incentives.

CONCLUSION

In this short review, it was shown how the current situation of feedstock availability, energy consumption and legislative pressure could allow the opportunity to turn three problems into one consolidated solution. The first aspect of this reality is the abundancy of renewable carbon in some locations on earth. It is, as was shown in this manuscript, the case of the province of Quebec but could be applied anywhere where forest or agricultural residues are available at large scale. Conversion of these residues to advanced transportation fuels could represent a very valuable option for this large source of carbon dioxide emissions around the world. The production of fuel (fermentation) or by other techniques such as thermochemical techniques unavoidably lead to the production of carbon dioxide. The latter is already mostly balanced (up to 86%) but using this CO_2 to produce methane, methanol or other fuel that could displace even more fossil fuel. One approach to do it would be to use excess renewable electricity to reduce carbon dioxide which will unavoidably be an energy-demanding process. Overall, such an approach would allow attacking GHG emissions on many aspects at the same time, combining three major challenges affecting our society and transforming it into a sole opportunity.

REFERENCES

[1] B. Pratima, "The Pulp and Paper Industry," in *The Pulp and Paper Industry - Emerging Waste Water Treatment Technologies*, vol. 27, no. 5, Elsevier Inc., 2017, pp. 9–29.

[2] de la faune et des parcs du Q. Ministère des forêts, "Compilation des données issues des registres forestiers 2017 Présentation faite à la Table de concertation sur le marché de la matière ligneuse," 2018.

[3] D. Bradley, "Canada report on bioenergy," Ottawa, 2010.

[4] S. Feng, S. Cheng, Z. Yuan, M. Leitch, and C. Xu, "Valorization of bark for chemicals and materials: A review," *Renew. Sustain. Energy Rev.*, vol. 26, pp. 560–578, 2013.

[5] NSERC, "Softwood barks, a green alternative for Quebec industries," *Dimensions*, 2013. [Online]. Available: http://www.nserc-crsng.gc.ca/ase-oro/Details-Detailles_eng.asp?id=641791.

[6] Fonds de Recherche Nature et Technologies du Québec, "Valorisation des lignines industrielles dans les nouveaux biocomposites," 2014. [Online]. Available: http://www.scientifique-en-chef.gouv.qc.ca/en/impacts-of-research-cat/valorisation-lignines-industrielles-nouveaux-biocomposites/.

[7] Hydro-Québec, "Comparaison des prix de l'électricité dans les grandes villes nord-américaines," 2018.

[8] Hydro-Québec, "Rate L - Industrial rate for large-power customers." .

[9] Hydro-Québec, "Rate D - Rate for residential and farm customers." .

[10] Hydro-Québec, "Photovoltaic solar power," 2018. [Online]. Available: http://www.hydroquebec.com/sustainable-development/documentation-center/solar-power.html.

[11] Natural Resources Canada, "Wind Energy," 2016. [Online]. Available: https://www.nrcan.gc.ca/energy/renewables/wind/7299.

[12] N. A. Ludin *et al.*, "Prospects of life cycle assessment of renewable energy from solar photovoltaic technologies: A review," *Renew. Sustain. Energy Rev.*, vol. 96, pp. 11–28, 2018.

[13] V. Khare, S. Nema, and P. Baredar, "Solar-wind hybrid renewable energy system: A review," *Renew. Sustain. Energy Rev.*, vol. 58, pp. 23–33, 2016.

[14] S. Quoilin, K. Kavvadias, A. Mercier, I. Pappone, and A. Zucker, "Quantifying self-consumption linked to solar home battery systems: Statistical analysis and economic assessment," *Appl. Energy*, vol. 182, pp. 58–67, 2016.

[15] S. Lee, D. Whaley, and W. Saman, "Electricity demand profile of Australian low energy houses," *Energy Procedia*, vol. 62, pp. 91–100, 2014.

[16] D. M. Gioutsos, K. Blok, L. van Velzen, and S. Moorman, "Cost-optimal electricity systems with increasing renewable energy penetration for islands across the globe," *Appl. Energy*, vol. 226, pp. 437–449, 2018.

[17] R. Loisel, L. Baranger, N. Chemouri, S. Spinu, and S. Pardo, "Economic evaluation of hybrid off-shore wind power and hydrogen storage system," *Int. J. Hydrogen Energy*, vol. 40, no. 21, pp. 6727–6739, 2015.

[18] M. Götz *et al.*, "Renewable Power-to-Gas: A technological and economic review," *Renew. Energy*, vol. 85, pp. 1371–1390, 2016.

[19] P. Schmidt, V. Batteiger, A. Roth, W. Weindorf, and T. Raksha, "Power-to-Liquids as Renewable Fuel Option for Aviation: A Review," *Chemie Ing. Tech.*, vol. 90, no. 1, pp. 127–140, 2018.

[20] Environment and Climate Change Canada, "Canadian Environmental Sustainability Indicators: Greenhouse Gas Emissions.," 2017.

[21] de l'environnement et de la lutte contre les changements climatiques Ministère du développement durable, "Inventaire québécois des émissions de gaz à effet de serre en 2015 et leur évolution depuis de 1990," Québec, 2018.

[22] N. Jacquet, G. Maniet, C. Vanderghem, F. Delvigne, and A. Richel, "Application of steam explosion as pretreatment on lignocellulosic material: A review," *Ind. Eng. Chem. Res.*, vol. 54, pp. 2593–2598, 2015.

[23] K. Zhang, Z. Pei, and D. Wang, "Organic solvent pretreatment of lignocellulosic biomass for biofuels and biochemicals: A review," *Bioresour. Technol.*, vol. 199, pp. 21–33, 2016.

[24] J. H. Lora and W. G. Glasser, "Recent industrial applications of lignin: A sustainable alternative to nonrenewable materials," 2002.

[25] P. Azadi, O. R. Inderwildi, R. Farnood, and D. A. King, "Liquid fuels, hydrogen and chemicals from lignin: A critical review," *Renew. Sustain. Energy Rev.*, vol. 21, pp. 506–523, 2013.

[26] Government of Canada, "Sales of fuel used for road motor vehicles, by province and territory," 2017. .

[27] Natural Resources Canada, "Ethanol," 2010. [Online]. Available: https://www.nrcan.gc.ca/energy/alternative-fuels/biofuels/3493. [Accessed: 14-Feb-2019].

[28] A. Ilas, P. Ralon, A. Rodriguez, and M. Taylor, "Renewable power generation costs in 2017," *Renew. Energy Agency, Int.*, pp. 0–161, 2018.

[29] T. Storch, T. Leukefeld, T. Fieback, and U. Gross, "Living houses with an energy-autonomy – results of monitoring," *Energy Procedia*, vol. 91, pp. 876–886, 2016.

[30] S.-Y. Chen, C.-Y. Chu, M.-J. Cheng, and C.-Y. Lin, "The Autonomous house: A bio-hydrogen based energy self-sufficient approach," *Int. J. Environ. Res. Public Heal.*, vol. 6, pp. 1515–1529, 2009.

[31] R. McKenna, E. Merkel, and W. Fichtner, "Energy autonomy in residential buildings: A techno-economic model-based analysis of the scale effects," *Appl. Energy*, vol. 189, pp. 800–815, Mar. 2017.

[32] O. Ellabban, H. Abu-Rub, and F. Blaabjerg, "Renewable energy resources: Current status, future prospects and their enabling technology," *Renew. Sustain. Energy Rev.*, vol. 39, pp. 748–764, 2014.

[33] M. Manwaring, D. Mursch, and K. Tilford, "Challenges and Opportunities For New Pumped Storage Development." NHA—Pumped Storage Development Council, p. 33, 2012.

[34] National Energy Board, "Market Snapshot: Pumped-storage hydro - the largest form of energy storage in Canada and a growing contributor to grid reliability," *National Energy Borad, Government of Canada*, 2016. [Online]. Available: https://www.neb-one.gc.ca/nrg/ntgrtd/mrkt/snpsht/2016/10-03pmpdstrghdr-eng.html.

[35] Z. Yang *et al.*, "Electrochemical Energy Storage for Green Grid," *Chem. Rev.*, vol. 111, no. 5, pp. 3577–3613, May 2011.

[36] F. V. Vázquez *et al.*, "Power-to-X technology using renewable electricity and carbon dioxide from ambient air: SOLETAIR proof-of-concept and improved process concept," *J. CO_2 Util.*, vol. 28, no. June, pp. 235–246, 2018.

[37] A. Sternberg and A. Bardow, "Power-to-What? – Environmental assessment of energy storage systems," *Energy Environ. Sci.*, vol. 8, no. 2, pp. 389–400, 2015.

[38] Audi, "Power-to-gas plant," 2013. [Online]. Available: https://www.audi.com.au/au/web/en/models/layer/technology/g-tron/power-to-gas-plant.html.

[39] Enbridge, "Blue Flame Bounty: A practical energy storage solution - Part 4: Power-to-Gas (P2G)," 2016. [Online]. Available: https://www.enbridge.com/energy-matters/energy-school/blue-flame-bounty-part-4.

[40] Carbon Recycling International, "Resource Efficiency by Carbon Recycling," 2006. [Online]. Available: http://www.carbonrecycling.is.

[41] Swiss Liquid Future AG, "Technology," 2016. [Online]. Available: http://www.swiss-liquid-future.ch/technology/?lang=en.

[42] X. Zhen and Y. Wang, "An overview of methanol as an internal combustion engine fuel," *Renew. Sustain. Energy Rev.*, vol. 52, pp. 477–493, 2017.

[43] J. Lane, "The rise, rise, rise of bio-methanol for fuels and chemical markets," *Biofuels Dig.*, 2018.

[44] L. Li, N. Zhao, W. Wei, and Y. Sun, "A review of research progress on CO_2 capture, storage, and utilization in Chinese Academy of Sciences," *Fuel*, vol. 108, pp. 112–130, 2013.

[45] X. Zhang, S. Yan, R. D. Tyagi, and R. Y. Surampalli, "Energy balance and greenhouse gas emissions of biodiesel production from oil derived from wastewater and wastewater sludge," *Renew. Energy*, vol. 55, pp. 392–403, 2013.

[46] M. Aresta, A. Dibenedetto, and A. Angelini, "The changing paradigm in CO_2 utilization," *J. CO_2 Util.*, vol. 3–4, pp. 65–73, 2013.

[47] MDDELCC, "Revenus des ventes aux enchères versés au Fonds vert," 2018. [Online]. Available: http://www.environnement.gouv.qc.ca/changements/carbone/revenus.htm.

[48] The Canadien Press, "Provinces have until the end of 2018 to submit carbon price plans: McKenna," 2017. .

[49] R. Slade, A. Bauen, and N. Shah, "The greenhouse gas emissions performance of cellulosic ethanol supply chains in Europe," *Biotechnol. Biofuels*, vol. 2, no. 1, p. 15, 2009.

MOVEMENT TOWARDS CIRCULAR ECONOMY AND E.U. WASTE MANAGEMENT REGULATION

Maria Loizidou, National Technical University of Athens

Stergios Vakalis, Free University of Bolzano & National Technical University of Athens

Konstantinos Moustakas, National Technical University of Athens

Dimitris Malamis, National Technical University of Athens

INTRODUCTION

The European Union (EU-27) has very active physical trade with the rest of the world with the exports increasing from 397 million tonnes to 568 million tonnes in 2011. At the same time, the imports increased from 1,340 million tonnes in 1999 to 1,629 million tonnes in 2011. Although the values indicate that the imports overwhelmingly outweigh the exports, the actual case is that biomass imports – exports are almost balanced and at the same time, the EU exports more manufacturing goods than it imports. The big imbalance is identified on the fuels and mineral products trade with more than 1,274 million tonnes being imported but only 215 million tonnes being exported. These aspects highlight the importance of substituting the existing linear economy by developing a circular economy that would reuse the materials and extend their lifetime. Circular economy can play a significant role in enhancing the resource efficiency by inserting sustainable practices like the eco-designing of products during manufacturing, along with post-use strategies like the 3Rs, i.e. reuse - recycling – recovery. On December 2nd, 2015, the EU approved an ambitious set of measures for Circular Economy: «Closing the loop». In the respect of resource efficiency, «Closing the loop» was an ambitious but promising package for advancing Circular Economy in the EU [1]. In this framework:

1. The value of products, materials, and resources is maintained in economy as much as possible
2. Waste generation is minimized
3. Economy and competitiveness are strengthened creating new business opportunities and introducing innovative products and services
4. Economic, social and environmental benefits are highlighted

These targets are expected to bring long-term benefits on several levels. It is estimated that 3.4 million new jobs will be created in the framework of circular economy, e.g. in the fields of: repairing, waste management, recycling, renting and leasing in the EU until 2030. The development of new jobs is expected to increase the GDP up to + 7%. In addition, the safety & stability for supplying raw/secondary materials will be enhanced. The overall savings are expected to be of up to 600 billion Euro, which corresponds to 8% of the annual turnover for companies in the EU. Another significant expected benefit is the reduction of the total annual emissions of greenhouse gases by 2-4%. The circular economy actions are promoted by international engagements and accords like the UN «Agenda 2030» - 17 Sustainable Development Goals (SDGs) and the Paris Agreement on Climate Change. These international agreements may enhance the activation of processes that speed up public and private investments in innovation especially in the fields of renewable energy and low carbon technologies.

WASTE MANAGEMENT AND CIRCULAR ECONOMY

In the framework of Circular Economy, waste management strategies aim to improve waste management in line with the EU waste hierarchy, address the existing implementation gaps and provide long-term vision and targets to guide investments. There are two key actions that will define the success of "Circular Economy" plans. Firstly, the Member States should improve their legislative schemes for waste management planning including the avoidance of overcapacity in residual waste treatment (incineration and mechanical-biological treatment). Secondly, the coherence between waste investments (under EU Cohesion Policy) and the waste hierarchy should be ensured. Following this narrative - and according to existing agreements - Member States should reduce MSW ending at landfills to 10% until 2030 as seen in Fig. 1 [2]. Also, they should increase the amount of packaging waste expected to be recycled and reused as seen in Table 1.

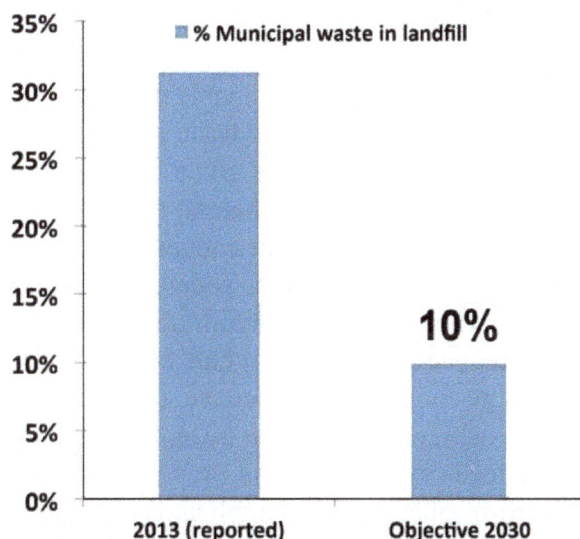

Fig. 1: Targeted reduction in waste landfilling by 2030

Table 1: New targets for recycling packaging waste

Packaging waste	Suggested Target 2025	Suggested Target 2030
Total	65%	75%
Plastics	55%	55%
Wood	60%	75%
Ferrous Metals	75%	85%
Aluminium	75%	85%
Glass	75%	85%
Paper-Cardboard	75%	85%

These ambitious goals can only be achieved by enhancing the existing reusing and recycling schemes by advancing the market for secondary raw materials. These actions would include the increase of use in recycled materials, nutrients and the reusing of treated wastewater. This parameter will be significantly improved by advancing the knowledge of material stocks and flows. Some identified key actions are the following:

1. Quality standards for secondary raw materials
2. EU regulation on fertilisers
3. Legislative proposal on minimum requirements for reused water
4. Analysis on the interface between chemicals, product, and waste legislation
5. EU-wide electronic system for cross-border transfers of waste

The previously mentioned points (nutrients, wastewater etc.) can be significantly advanced by means of separate and targeted collection and valorization of biomass and biowaste for the production of bioproducts. The scope is to support an efficient use of wood and bio-based products and increase the recycling of biowaste. The efficient use of bio-based resources should be promoted through a series of measures, including the cascading use of biomass and the support of bio-economy innovation. EU biowaste management policy aims to reduce the environmental and health impacts of biowaste and improve Europe's resource efficiency. The long-term goal is to turn Europe into a recycling society, avoiding waste and using unavoidable waste as a resource wherever possible. The aim is to achieve much higher levels of recycling and to minimize the extraction of additional natural resources. Proper waste management is a key element in ensuring resource efficiency and the sustainable growth of European economies.

Food waste has been the center of attention as a basic subcategory of biowaste. The EU has set the target of 20% food waste recycling and has given an extension to 7 Member States for achieving recycling targets higher than 20%. Nonetheless, a revision of these targets is expected to be examined in 2025 for setting higher targets. In a legislative proposal, the reduction of food waste should reach the level of at least 30% by 2025. Today, approximately 100 million tonnes of food are wasted every year in the EU and a main objective is to reach the Sustainable Development Goal (SDG) and reduce food waste in half by 2030. For this purpose, EU should develop a methodology to measure food waste and create a platform for the SDG on food waste where the best practices and the achieved results can be shared. A further clarification of the EU legislation on waste, food and feed is necessary. As presented in the previous paragraph, food waste has been an underutilized resource and the application of best practices will significantly promote Circular Economy for waste management. This study briefly presents two characteristic EU funded projects that have been specifically developed for promoting the collection and valorization of food waste.

EU LIFE + PROJECTS "DRYWASTE" AND "WASTE2BIO"

The first presented project is the LIFE+ "DRYWASTE" project (LIFE08 ENV/GR/000566) that targeted the development and demonstration of an innovative household dryer for the treatment of organic waste. In addition, the recovered and processed material was of high purity and exceeded 99%. This high value can be attributed to the fact that the participants underwent extensive training. The average operational characteristics of prototype domestic dryers that were used in the framework of the project are presented in Table 2 [3].

Table 2: Energy consumption of the prototype system operating for 8hr and energy cost at 0.0709 €/kWh (Samples mass = 0.5 kg)

Drying temperature (°C)	Daily Energy consumption (kWh)	Monthly Energy consumption (kWh)	Monthly operational cost (€/month)
60	0.73	22.00	1.56
70	0.93	28.00	1.99
80	1.13	34.00	2.40

Food waste has typical moisture content higher than 70 -75% and the removal of the water reduces significantly the mass and the volume of the waste [4]. The average mass reduction after drying was measured to be from 58.31 to 76.81% w/w and in most cases, the moisture content was well below 10%. At the same time, drying sterilizes the waste and thus facilitates long-term storage solutions. On the one hand, the dried material was found to have very interesting physicochemical characteristics like high carbon content and high sugar content. As a result, the dried food waste was used as raw material for biochar [5] production and particleboards production [6] respectively, with both products having very good characteristics.

On the other hand, the dried material has very high carbohydrate and cellulose contents, i.e. 23-24 % w/w and 17.2 % w/w respectively. Drying at the source tends to preserve the non-structural carbohydrates and this was a good indicator that the dried biowaste could be essential in boosting the bioconversion of bio-waste to ethanol [7]. Thus, the EU LIFE+ project "Waste2Bio - Development and demonstration of an innovative method of converting waste into bioethanol" was developed. As mentioned in the objectives of the project, the overall goal of the Waste2Bio project was to develop a method of converting biowaste into bioethanol fuel." For this purpose, a pilot unit was developed and operated at the Unit of Environmental Science and Technology of NTUA and is presented in Fig. 2 [8].

The reactor consists of a pre-treatment and a main reactor. The dried biowaste enters the pre-treatment reactor along with water and enzymes. The enzymatic hydrolysis of the biowaste was correlated to the substrate concentration of enzymes, at the level of pretreatment and at the potential presence of sulfuric acid. The scope was to produce bioethanol via the degradation of polysaccharides and a subsequent simultaneous saccharification and fermentation process in the main reactor. Saccharification and Fermentation (SSF) under batch conditions produced ethanol values of 30.32 g/L (pretreatment: 1000C for 1h & sulfuric acid). In this case, the substrate concentration was 30% and in fed-batch conditions with less substrate concentration of 40% and less enzyme loading (7FPU/g vs 10FPU/g) producing higher amounts of ethanol, i.e. 42.66 g/L. Combination of SHF/SSF processes returned even higher values between 34.85 g/L and 42.78 g/L.

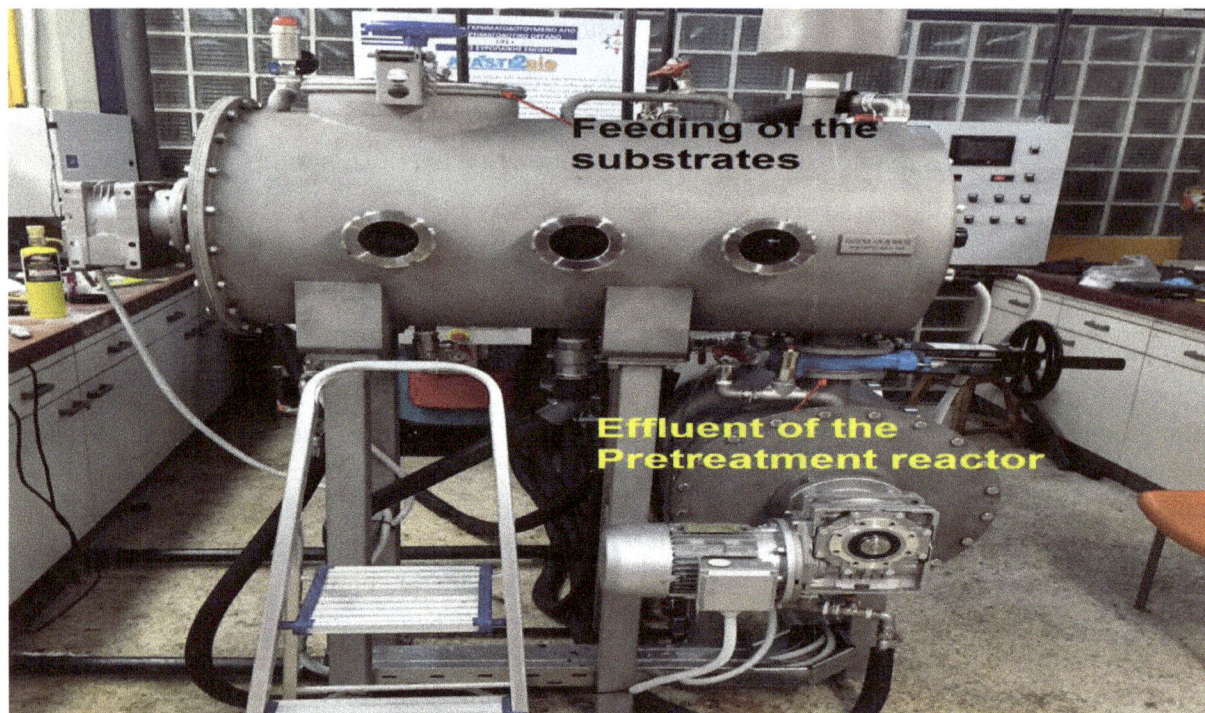
Fig. 2: A prototype pilot-scale reactor for bioethanol production

DISCUSSION AND SUGGESTIONS

Overall, circular economy provides a good framework for increasing the sustainability of supply chains. On the consumption side, the actions for encouraging reuse activities (e.g. waste proposal), the addition of Circular Economy criteria in Green Public Procurement, the initiation of independent testing programs for assessing planned obsolescence, are key for avoiding waste generation and for providing consumers with reliable information on environmental impact of products. Overall, the goal should be the creation of the right environment for innovation & investment. Several strategies can enhance the role of circular economy in the field of food waste management, such as the recyclability in eco-design, the optimization of best practices for waste management & resource efficiency in industrial sectors and the application of Industrial Symbiosis. This study presented two successful (EU funded) food waste management schemes that promote valorization pathways for materials and fuels production.

REFERENCES

[1] European Commission, 2015. COMMUNICATION FROM THE COMMISSION TO THE EUROPEAN PARLIAMENT, THE COUNCIL, THE EUROPEAN ECONOMIC AND SOCIAL COMMITTEE AND THE COMMITTEE OF THE REGIONS Closing the loop - An EU action plan for the Circular Economy COM/2015/0614 final

[2] European Commission, 2018. DIRECTIVE OF THE EUROPEAN PARLIAMENT AND OF THE COUNCIL on the reduction of the impact of certain plastic products on the environment. Brussels, 28.5.2018 COM (2018) 340 final

[3] Development and demonstration of an innovative household dryer for the treatment of organic waste, DRYWASTE, European Commission, LIFE+, LIFE08 ENV/GR/000566, 2010 – 2012

[4] S. Vakalis, A. Sotiropoulos, K. Moustakas, D. Malamis, K. Vekkos, M. Baratieri (2016). Characterization of Hotel Bio-waste by Means of Simultaneous Thermal Analysis. Waste and Biomass Valorization 7; 649-657.

[5] S. Vakalis, D. Semitekolos, J. Novakovic, K. Moustakas, L. Zoumpoulakis and M. Loizidou (2018). Introduction to the concept of particleboard production from mixtures of sawdust and dried food waste. Waste and Biomass Valorization. DOI: 10.1007/s12649-018-0214-0.

[6] S. Vakalis, A. Sotiropoulos, K. Moustakas, D. Malamis, K. Vekkos, M. Baratieri (2017). Thermochemical valorization and characterization of household bio-waste. Journal of Environmental Management 203; 648-654.

[7] A. Sotiropoulos, I. Vourka, A. Erotokritou, J. Novakovic, K. Valta, V Panaretou, S. Vakalis, T. Thanos, K. Moustakas, D. Malamis (2016). Combination of decentralized waste drying and SSF techniques for household biowaste minimization and ethanol production. Waste Management 52; 353 – 359.

[8] Development and demonstration of an innovative method of converting waste into bioethanol, European Commission, Waste2Bio, LIFE+, LIFE11 ENV/GR/000949, 2012-2016

APPLICATIONS OF THE 3T METHOD AS AN EFFICIENCY ASSESSMENT TOOL FOR WASTE-TO-ENERGY FACILITIES AND NUMERICAL COMPARISONS WITH THE R1 FORMULA

Stergios Vakalis, Free University of Bolzano
& National Technical University of Athens

ABSTRACT

The scope of the work is to present an integrated method for assessing the energy efficiency of waste-to-energy plants. A characteristic aspect of waste-to-energy plants is that they are not only energy generation facilities but also metal recovery operations. Thus, the efficiency formulas should take into consideration not only the Combined Heat and Power (CHP) production but also the quality of the materials. In addition, the gradual commercialization of novel technologies like gasification and pyrolysis is making the inclusion of materials a necessity. This work presents numerical solutions between the 3T method and the R1 formula and assesses the limitations and constraints of each methodology

INTRODUCTION

The Waste Framework Directive 2008/98/EU separates the waste management strategies into Recovery Operations and Disposal Operations (ANNEXES I and II respectively). Waste-to-energy technologies that use waste as fuel for efficient energy generation are considered to be "Recovery Operations", i.e. R 1. When the scope is the destruction/ reduction of the waste before landfilling, the waste-to-energy technologies are considered to be "Disposal Operations", i.e. D 10. This "issue of duality" for the case of waste-to-energy plants, which was described above, created the necessity for the establishment of a set of criteria that would distinguish the plants that were efficient enough in order to be considered as "Recovery Operation" plants. This issue has been of high importance because each waste-to-energy facility can potentially fall into both categories according to the assessment tool that sets the bar. In order to address this issue the European Commission integrated the R1 formula into the second revision of the Waste Framework Directive of 2008 [1] and improved it in the Directive 2015/1127/EU that entered into force from July 31st, 2016 [2].

It should be stated for the record, that the R1 formula has been a very helpful tool for assessing waste-to-energy plants and played an important role in defining the general framework. The problems with the use of the R1 formula start from the fact that the formula is not thermodynamically consistent since it is self-proclaimed to be more a "utilization efficiency" formula rather than a pure energy efficiency formula. In addition, modern technologies that treat thermally municipal solid waste are polygeneration facilities because they produce several output streams. Such typical cases are the waste gasification facilities that produce several streams in addition to Combined Heat and Power (CHP) production like char or even reformed fuels like methanol. In addition, waste-to-energy facilities are effectively metal recovery units. These parameters are not taken into consideration by the R1 formula, which primarily deals with the case of conventional incineration for CHP production [3]. This work applies the 3T method, which was introduced by Vakalis et al. [4] and assesses the applicability of the method as an alternative to the R1 energy efficiency formula. The scope of this study is to present the numerical results from the analysis of waste-to-energy plants, both with the R1 formula and with the 3T method and to assess the findings of this analysis.

CALCULATION PARAMETERS FOR THE 3T METHOD

Vakalis et al. [4] took into consideration the issues that were raised about the R1 formula. The authors provided an alternative methodology that they name "the 3T method". The four parameters that are taken into consideration are: CHP efficiency, CHP exergy efficiency, exergy efficiency of products and exergy efficiency of recovered metals. These four parameters are combined in a radar graph as shown in Fig. 1 (left) and the area of the trapezoid is calculated. This calculation provides a unique value that is characteristic of each plant by taking into consideration not only the CHP production but also the quality of the produced materials. Thus, different waste-to-energy technologies, like combustion or gasification, can now be compared directly. As shown in Fig. 1 (right) a more simplified solution can be developed for the specific case of incineration.

It has to be stated that the methods do not have a same value system and the final efficiencies that they calculate are not comparable. For the case of the R1 formula, the facilities that attain values over 0.65 (or 0.6 for older plants) achieve the R1 status. It is usually the case that efficient co-generation plants can achieve significantly higher values than the values reported above. On the other hand, conventional waste-to-energy plants achieve values of 0.2 – 0.3 with the 3T method. The comparison of the methods becomes very interesting for the case of gasification or pyrolysis plants where the R1 formula is not able to take into consideration other final products except the CHP, like char or bio-oil.

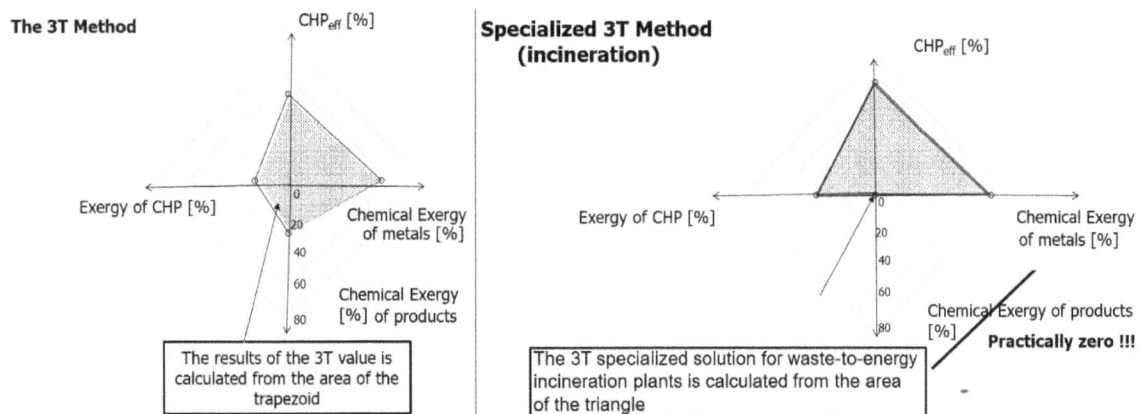

Fig 1: The generalized 3T method (left) and the specialized solution for the case of incineration (right)

RESULTS AND NUMERICAL COMPARISONS

Three characteristic plants were used for applying the R1 formula and the 3T method in order to analyze the comparative results. The parameters of three characteristic plants, represented as A, B & C and shown in Table 1 have been used for the calculation of the 3T method by Vakalis et al. [4].

Table 1: Parameters of representative plants for the calculation of R1 formula & 3T method

	Plant A	Plant B	Plant C
Electrical efficiency [%]	17	21	27
Thermal efficiency [%]	55	45	45
Temperature of output heat [°C]	85	85	85
Physical exergy efficiency [%]	25.22	27.46	33.23
Exergy efficiency of metals [%]	35	35	35

The results of the calculations are shown in Fig. 2. The R1 returns the values of 1.07, 1.07 and 1.23 respectively. The 3T values for the same plants are 0.217, 0.206 and 0.246. Plants A & B have the same R1 values. For the same cases, the 3T method provides vastly different results since the range of returned results from the application of the 3T method has been shown to be between 0.2 and 0.3 for most CHP technologies. It has to be noted that for some non-conventional Energy-from-Waste technologies the 3T value have been reported to slightly deviate from the previously mentioned range [5]. The discrepancy in the results is such that it becomes clear that the two methods take into consideration different things. The recovery of metals is (for example) one significant parameter not considered in the R1 formula.

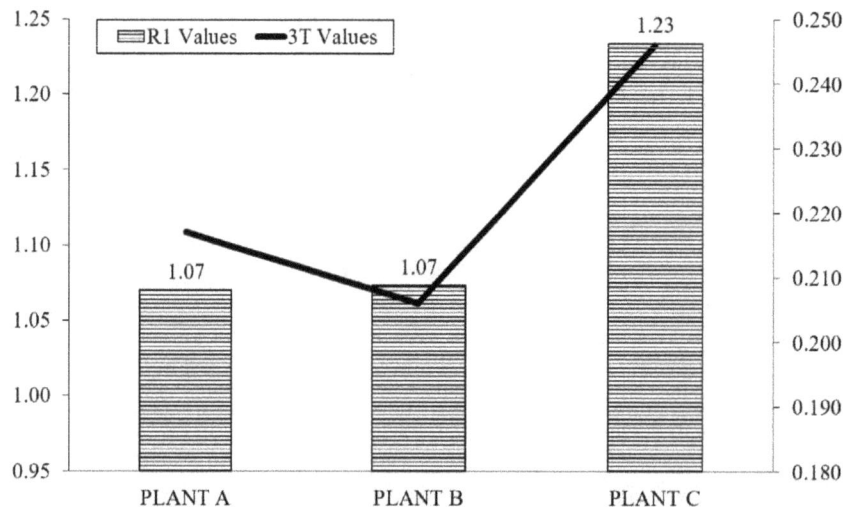

Fig 2: Results of applying the R1 formula and the 3T method

Fig. 3 presents the results of applying the R1 formula and the 3T method for plants of 15 % electrical efficiency and increasing heat efficiencies (X-axis). A single line projects the results from the R1 formula, since the result is the same for any percentage of material recovery. The results in Fig. 3 show that the aspect (and percentage) of metal recovery is being reflected in the results from the 3T method but not in the R1 formula. Contrary to the R1 formula, the 3T method returns different results for different percentages of metal recovery. Again, it should be mentioned that the two methods have different value systems return results within different ranges.

On a second level, the 3T method uses the exergy of CHP for the calculation of the 3T values and this is directly correlated with the temperature (and pressure) of the produced steam. This parameter again cannot be considered by the R1 formula, although it should be mentioned that the enthalpy of steam is considered.

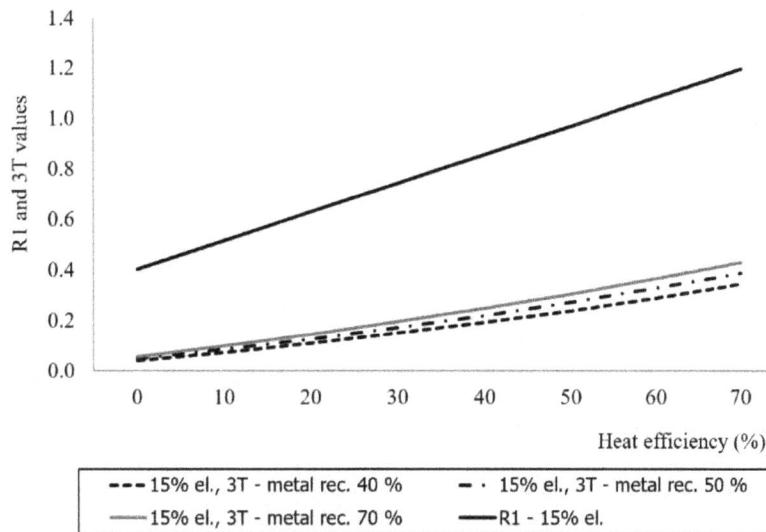

Fig. 3: Comparison between R1 and 3T values for different heat efficiencies and the effect of metal recovery

REFERENCES

[1] European Union (2008) Directive 2008/98/EC of the European Parliament and of the Council of 19 November 2008 on waste and repealing certain Directives. Off. J. Eur. Union L 312, 3–30.

[2] European Commission (2015) COMMISSION DIRECTIVE (EU) 2015/1127 of 10 July 2015 amending Annex II to Directive 2008/98/EC of the European Parliament and of the Council on waste and repealing certain Directives.   July 2015, Brussels, Belgium.

[3] European Commission, DG (Directorate-General) Environment (2011) Guidelines on the interpretation of the R1 energy efficiency formula for incineration facilities dedicated to the processing of municipal solid waste according to annex II of Directive 2008/98/EC on waste.

[4] S. Vakalis, K. Moustakas and M. Loizidou (2018). Assessing the 3T method as a replacement to R1 formula for measuring the efficiency of waste-to-energy plants. Waste Management & Research 36, 810 – 817.

[5] S. Vakalis and K. Moustakas (2019). Applications of the 3T Method and the R1 Formula as Efficiency Assessment Tools for Comparing Waste-to-Energy and Landfilling. Energies 2019, 12, 1066.

NEED FOR SUSTAINABLE WASTE MANAGEMENT IN COLOMBIA

Enrique Posada Restrepo, WtERT Council - Colombia

Gabriel Naranjo Pizano, WtERT – Colombia

Walter Ospina, WtERT Council - Colombia

ABSTRACT

Colombia is a country of 46 million people located in South America, with special historical, geographical, demographic, and cultural antecedents that play an important role in the genesis of many problems, such as waste handling. Although there is a clear need for sustainable waste management, there are important barriers. Some of them are considered and discussed in relation to options for developing Waste to Energy (WTE) projects. The Colombian electricity companies seem to consider WTE projects that produce electricity only from the point of view of comparing investment costs per installed kilowatt and the cost of generation per kilowatt-hour, to hydroelectric, wind or solar projects. In general, this means that the generation of electricity from MSW appears as neither profitable nor viable. This conclusion is also felt in the companies and entities of the waste sector. Of course, in this way, the integral, comprehensive picture is not included, as the waste sector must consider a more sustainable and broader perspective and see electricity generation as one aspect of WTE and not the only goal. The interests of the electricity sector and the waste management sector, apparently, are not the same and do not seem to coincide. Sustainability is a common ground to consider and also the need to integrate solutions looking for the real (evident and hidden) cost and benefits to society. Here, the authors present a model of application of WTE to solid waste in Colombia to help provide some perspective on this matter.

INTRODUCTION

Put introduction here. Put introduction here. Put introduction here. Put introduction here. Put introduction here. Put introduction here. Put introduction here. Put introduction here. Put introduction here. Put introduction here. Put introduction here.

Colombia is the third-most populous country in Latin America, with around 46 million people in 2017. Its population is mostly concentrated in the Andean highlands and along the Caribbean coast, with 31 cities of more than 200.000 inhabitants and 65 with more than 100.000.the capital Bogotá has 8.2 million inhabitants; Medellin and Cali have 2.5 million, and there are around 10 other cities with more than 500.000 inhabitants. Formerly a rural society, Colombia is now one of the most urbanized countries in the region, with and an urban population estimated at 76%. The average national unemployment rate in 2017 was 9.4%, but informality is a big problem for the labor market. The Gross domestic product per capita in 2017 was 6.472 U.S. dollars. As in 2016, the National Administrative Department of Statistics (DANE) reported that 28.0% of the population was below the poverty line, and of those, 8.5% were in extreme poverty.

High levels of poverty, indigence, and unemployment, are associated with informal waste recycling practices, especially among the poor. Around 20 people per 10000 inhabitants in Colombia work in recycling tasks, 3.9 of those are organized in collective groups; 5.7 are organized as workers in separating plants; 14.5 are associated to the recollection systems; and 9.7 are in other places. Fig. 1, prepared according to date from an OPS (Pan-American Health Organization) study [1] compares the daily per capita urban solid waste generation in Colombia to the situation in Latin America, showing that Colombia generates waste at a rate a little lower than the average. With a medium generation of 0.54 kg/inhab./day, the estimated daily generation will be around 26.000 tons. Colombia has a very complete waste recollection system, reaching 100 % of the total generated according to this study. This figure seems to be somewhat unrealistic, as there are instances in which people still throw solid waste to streams and rivers.

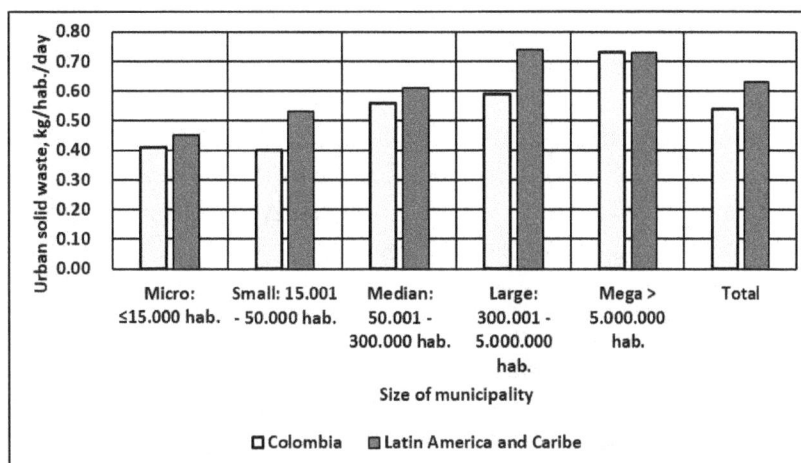

Fig. 1: Daily per capita urban solid waste generation in Colombia and Latin America-Caribe countries according to size of municipality

In most Colombian cities, informal workers walk the streets recovering valuable waste (paper, cardboard, metal, glass, plastics) from the bags and recipients that people put outside their homes for municipal waste collection. These workers take the materials to more or less well-organized places, situated in some zones of the cities, where they sell the materials to more sophisticated dealers, who consolidate and pack them for sale to industrial waste material users. In the case of industrial and institutional generators of municipal waste, and even for some residential complexes, it is common that cooperatives or specialized firms separate and recycle valuable waste, giving opportunities for better working conditions to recycling operators. In Colombia, recycling has been in practice for many years, and it employs about 100.000 people, with about 30% associated to 128 cooperatives that even have a National Association of Recyclers (ANR), and 70% working independently. This makes Colombia the nation in Latin American with the highest proportional number of recyclers. The relative improvement in the organization of these people has been also favored by the existence of a regulatory framework that recognizes them as actors with the ability to assume legal and institutional commitments. Colombia is a model in the region in the recycling of paper and cardboard, with a recovery of 57%. This has to do with the existence of industrial plants that are able to use these materials in their process, which has favored a well-organized recycling scheme. Currently [3], the recycling rate of waste such as paper, cardboard, glass, metal and plastics is 17%, and by 2018 the goal was to achieve a recycling target of 20% as a result of the implementation of regulatory instruments in the regulation of the public cleaning service and the tariff frameworks, processes that the national government is working on. The rest of the waste goes to waste dumps or sanitary landfills, as there is no thermal treatment or WTE facilities in the country. Following the OPS report, 81.8 % of this goes to sanitary landfills (100 % for the case of the large cities); 4.1 % to controlled dumps, 12.5% to open dumps, 1.3 % to open field burning and 0.3 % to rivers and water bodies [1]. Out of these disposing facilities, very few have lixiviated treating plants or methane burning systems. Space is becoming an issue and there are growing concerns and limitations about the growth of the landfill areas in the coming years. In other cases, environmental concerns are becoming more and more important [1], [4], [5], [6].

LEGAL FRAMEWORK IN COLOMBIA FOR THE INTEGRAL AND SUSTAINABLE MANAGEMENT SYSTEM OF SOLID MUNICIPAL WASTE (MSW) AND THE USE OF WTE SYSTEMS

In Colombia, general planning is carried by the National Direction of Planning, through directives generated by COMPES (Consejo Nacional de Política Económica y Social, National Council of Economic and Social Policy). There is currently a defined National Policy for the Management of Solid Waste, updated through the CONPES 3874 of 2016 [4]. The policy is based on:

- Reduce the generation of MSW (the same as USR, urban solid residue).
- Minimize MSW that are deposited in landfills or dumps as a final disposal.
- Promote the reuse, use and treatment of solid waste.
- Minimize and avoid the generation of greenhouse gases

In the promotion of treatment and use of wastes, there is a clear space for the application of energy recovery, waste to fuel and waste to product technologies and measures. In addition, there is the possibility of actions for the promotion of biological treatment of the different fractions that make up the MSW, as well as actions for the generation of energy from unusable waste. Table 1 shows the principles of circular economy adopted by CONPES [4].

The ways of proceeding described in table 1 imply, in many cases, new ideas and beliefs, as compared to the usual practices [11],[13]. To establish those mindsets among people (consumers, producers, government, leaders, and citizens) it is necessary to work out ideas about change, points of view, creativity, testing, observing, imagining, proposing, normalizing, developing culture, and education. To attain changes, there is the need of investing time, resources, and developing tools, good communication and leadership. Part of the new mindset, is to have a far-reaching and constructive interpretation of the terms associated to the definitions. In principle, the norms and laws try to define everything as much as possible, but still there must be room to widen the definitions in such a way that sustainability (economics, social impact, ecology and environmental impact) is favored. For example, use and valorization (or treatment) have ample meanings. One of them (following Colombian norm CRA 720-2015, [5]), defines use as the activity that recyclers do on the streets by themselves or as organized groups, or the activity of companies that collect MSW and have selective routes to collect recyclable waste. In the context of this norm, the definitions are presented to establish limits and conditions for waste collecting tariffs. However, when applied to a biogas plant, they refer to the biological treatment of the organic fraction to obtain a soil conditioner and to the energy recovery from the biogas generated, which is a biofuel with various types of uses.

Table 1: Principles of Circular Economy

Principle	Description
Eco-conception	Considers environmental impacts throughout the life cycle of a product and integrates them from the very conception of the process
Industrial and territorial ecology	Establishes a mode of industrial organization in the territory, characterized by an optimized management of stocks and flows of materials, energy, and services
Economics of functionality	Privileges the use of goods over their possession, and the sale of services over the sale of goods
Second use	Reintroduces products that no longer correspond to the initial needs of consumers into the economic circuit.
Re-use	Re-uses certain waste or certain parts of it that can still be used to manufacture new products.
Repair	Finds new life for damaged products.
Using	Takes advantage of the materials found in waste.
Valorization (treatment)	Energetically and functionally makes good use of waste that cannot be recycled.

It is important to explain that the mentioned Colombian regulations, so far, apply to the inorganic fraction of the MSW and do not deal with the organic fraction, nor to waste to energy or to other more sophisticated waste to use applications. For these cases, the expedition of norms for its use is still pending, as mentioned in a 2016 Report of the Superintendence of Residential Public Services [6]. There it is called for generating and enabling institutional instruments that promote biological treatment of organic waste, coordinating necessary measures for the generation of energy from non-usable waste, and establishing measures for the use of construction debris.

Despite this, and in accordance with the definition of recycling adopted in the mentioned CONPES document, "Process of physical or chemical or biological transformation of the materials from the potentially usable waste, for its reincorporation in the productive cycle", there is ample space to accommodate in it most valorization and treatment processes.

Apart from the national policies and norms, Colombia has assumed international compromises that have to do with the management, use and valorization of the MSW. This is the case of COP21-2015 where the country has agreed to reduce 20 % of its GHG emissions in 2030. Since the MSW sector of Colombia is the fourth largest contributor to the generation of such gases [7], WTE projects, such as a biogas plant, prevent methane, an important GHG, from escaping into the atmosphere. With the entry of Colombia into the OECD (Organization for Economic Co-operation and Development) becoming a reality, the Environmental Policy

Committee of this body requires that, in the matter of non-hazardous solid waste, Colombia should [8] have comprehensive waste management policies that meet environmental protection objectives, taking into account economic constraints and local conditions and also economically efficient and environmentally reasonable waste management.

In the United Nations 2015 summit, Colombia subscribed to the 17 sustainable development objectives, which went into application in 2016. Specifically, objectives 11 and 12, refer to the integral and sustainable management of the MSW, where WTE projects, like a biogas plant and incineration systems, apply perfectly.

In the National Development Plan of Colombia for the period of 2014-2018, called "All for a New Country" [9], it is proposed to issue a general law for the integral management of solid waste, to harmonize the existing regulations. The National Development Plan refers to desirable efficiencies in the integrated management of solid waste for local entities, which will allow for incentives when the Plans for Integral Solid Waste Management (called PGIRS) include projects for using and taking advantage of waste. It mentions the integral management of solid waste and calls for establishing policies and regulations aimed at strengthening management through regional models that generate economies of scale and encourage investment to ensure adequate systems for the final disposal and utilization. This includes mechanisms for definition, coordination and articulation of inter-sectoral policy, plans and programs for the integral management of solid waste in the country, including incentives for their use.

Decree 1077 of 2015 of the Ministry of the Environment regulates the provision of the waste-handling services and requires that the municipalities incorporate their use in the PGIRS [10]. It calls for the development of projects for the use of waste to attain its incorporation into the productive cycle with social, economic and financial viability that guarantees their sustainability over time and their evaluation through the establishment of goals by the municipality or district. In the development of this activity, they should give priority to feasibility studies on waste use. The decree goes all the way to state the duty of the territorial entity to incorporate the technical and operational conditions that allow the development of schemes of use and valorization of waste.

In article 361 of the Political Constitution of Colombia, 1991, it is indicated that the government must ensure the provision of public services to all the inhabitants of the country, while Law 142 of 1994, allows the municipalities to delegate this work in authorized private entities. Therefore, the Municipalities of Colombia can be directly promoters and/or owners of WTE plants or can be assigned to an existing entity for that purpose.

It follows that, according to the legal framework described, the execution of WTE projects in Colombia is possible from the normative point of view.

EXISTING BARRIERS FOR WTE IN COLOMBIA AND SUGGESTIONS TO MANAGE THEM

The authors, through their experience as promoters of WTE projects; members of WTERT - Colombia council [14]; as project engineers and interested subjects in the field and participants in several international and national forums, seminars, university courses and conferences, have been able to perceive a list of barriers. They make the development of WTE projects in Colombia somewhat difficult, at least in the short term. They are listed and analyzed in Table 2. For each one, suggestions are proposed for their managing and mitigation, with the purpose of promoting the realization of sustainable and well-designed projects of WTE systems that become truly beneficial for the country.

Table 2: Barriers to WTE implementation and proposed actions to overcome them

Barrier	Proposed actions
Lack of knowledge of WTE technologies	Encourage knowledge and public awareness among people, academy, designers, consulting firms, authorities, public officials, companies, and entities responsible for managing waste.
Lack of an appropriate technological base and lack of complete engineering when doing projects	<ul><li>Stimulate local technology and engineering in the projects to develop technology, and create jobs and prosperity.</li><li>Promote application and knowledge of all the engineering stages (conceptual, basic, detailed, execution) to each project. [12]</li><li>Planning and design based on the establishment of clear objectives.</li><li>Execution under technical criteria.</li><li>Control and monitoring of execution to be within the budgeted costs and with the required quality.</li><li>Exercise feedback and recurrent work, based on discipline, interdisciplinary group work, motivation and leadership, to achieve constant perfection</li></ul>
The interests and influence of existing waste managing concessionaires	<ul><li>Stimulate the existing concessionaries to become part of the new WTE schemes, especially as investors, considering the large capital requirements for these projects.</li><li>Introduce the WTE projects as a growing part of the total solution, as it would be quite difficult to change the existing systems at once.</li><li>Allow WTE systems to generate new opportunities and new areas of work for concessionaries.</li></ul>
Political-technological relationships	<ul><li>Stimulate strong connections between engineering (represented by professional societies and guilds) and politicians and policy makers.</li><li>Professional societies and groups should reach maturity and develop capabilities to study and propose WTE projects and alternatives, to exercise good communication, and to be able to convince and be listened to .</li></ul>
Required initial investments	<ul><li>Develop procedures to demonstrate the advantages and real operating benefits, costs, and initial investments for a WTE project from the technical, environmental, and social points of view,.</li><li>Include critical analysis of the real costs, associated investments, and life-cycle environmental impacts of sanitary landfills. This could help to better define the MSWH projects.</li></ul>
Legal framework for WTE projects	<ul><li>Stimulate the completing of the regulatory norms for the existing laws</li><li>Try to stimulate simplifying the regulatory process following the already existing developments on WTE in several countries.</li></ul>

Barrier	Proposed actions
Interests and perspectives of WTE as seen from the electricity sector	• The waste sector must consider a broader perspective and see electricity generation as one aspect of WTE and not the only goal. • It is necessary to understand the WTE projects from the economic point of view and add to the economic considerations the corresponding social and environmental evident and hidden benefits, which should be monetized as much as possible and included in the cash flow to evaluate a given project. • To get some initial experience, it seems advisable, to work out WTE projects in Colombia, with low electricity production
Environmental permits	• WTE projects should be considered part of the waste sector and not part of the Colombian electricity sector for environmental permit considerations. • Devise integral evaluating protocols that favor the development of these systems, and consider the importance of the environmental benefits they have
Availability of land	• Make use of the advantage that WTE projects have in relation to land use, as compared to sanitary landfills
Tipping or gate fees	Some revisions of this should be proposed to help finance a WTE project, understanding, however, that gate fees will not likely be the main source of financing these projects, which require high investments and public equity.
Zero waste policies	There must be knowledge and practical wisdom, to understand the ways of reaching minimum waste generation and minimum waste disposed improperly into the environment, as absolute Zero waste is impossible. WTE is perhaps the best practical advance in the right direction

PROPOSED MODEL FOR APPLICATION OF WTE TO SOLID WASTE IN COLOMBIA

To help provide some perspective on this matter, the authors present here a model of application of WTE to solid waste in Colombia. It is based on simulating the impact on the economy of these projects, according to variations on the auxiliary fuel used (natural gas or coal) and in the water content of the treated waste (which varies according to the percentage of organic domestic biodegradable material separated before the WTE facility).

The theoretical model has been applied to the specific case of municipal waste for the city of Medellin. Basic information for this is the compositions of the MSW and of the coal and natural gas to be used as co-combustion fuels, plus their heat powers. Two extreme cases are considered for the waste. The first one, waste as currently generated, considers the average quality of the MSW in the city of Medellin, which is quite rich in organic materials and so, very high in water content. In the second case, previously separated waste is considered, removing 75 % of organic, 50 % of paper, 20 % of plastics, 55 % of glass, 60 % of cardboard and 50 % of metals of the generated waste. This would amount to a 45 % of the initial as generated MSW. Intermediate cases are also considered to study effects of percentage of domestic organics separated and, at the same time, the effect of water waste content. For each case the waste heat power was calculated based on waste composition.

Fig. 2 shows waste heat power and water content for the range of cases considered.

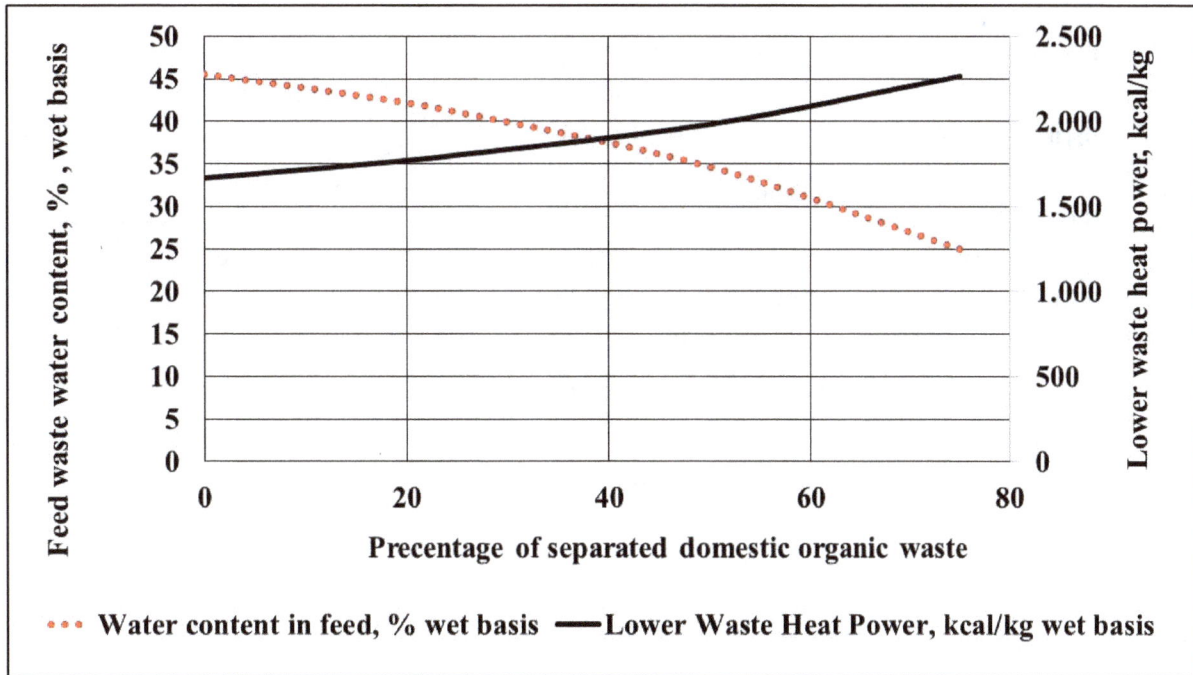

Fig. 2: Waste heat power and water content for the range of cases considered in the model

An iterative model calculation was developed for each case, using the solver routine of MS excel in which the amount of co-combustion fuel as a fraction of waste was iterated until the expected convergence was found, considering species mass balance and energy balance in the burner furnace and in the steam generator. Adiabatic gas temperature after combustion was taken at 1700 Celsius degrees. Steam energy conversion to electricity was taken as 30 %. Furnace combustion was simulated at 20 % air excess for waste, 10% air excess for natural gas and 30% air excess for coal.

Fig. 3 shows the proportion of auxiliary fuel used for the range of cases considered, both with natural gas and coal co-combustion.

The model estimates required electricity sale prices and electric power generated for a unit capable of treating 500 ton/day of waste after recycling at current levels and organic waste separation at the simulated levels. It also estimates number of processing plants and total generated electricity for the treatment of all the separated waste in WTE. Fig. 4 shows the expected results for the model.

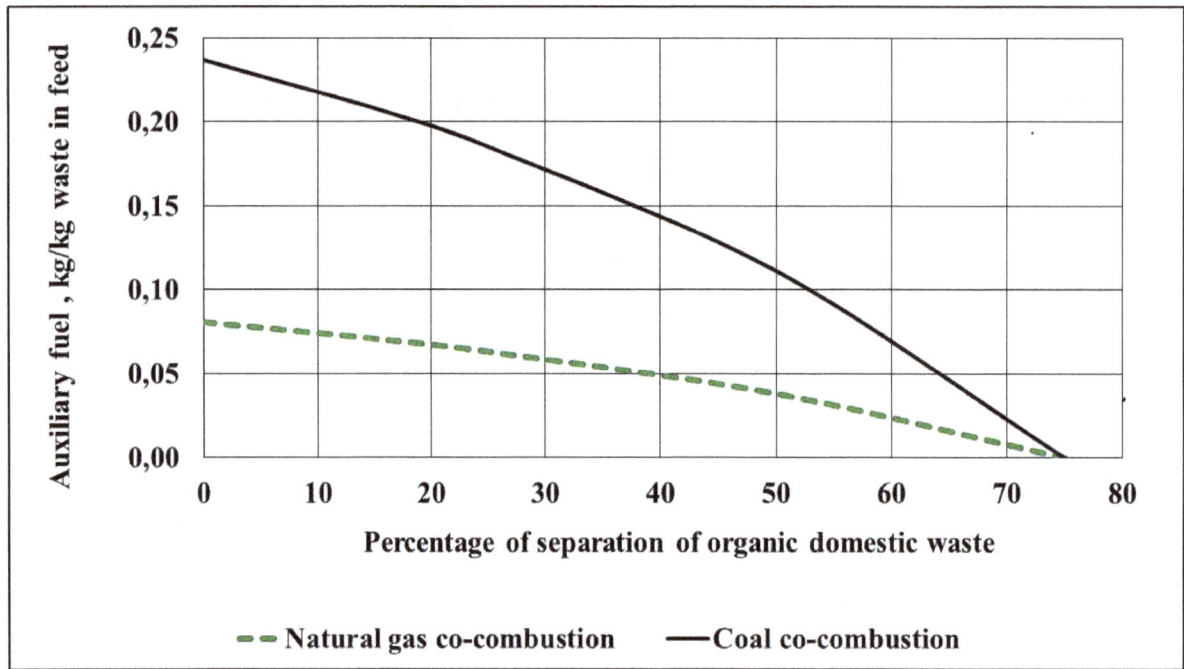

Fig. 3: Auxiliary fuel used for the range of cases considered

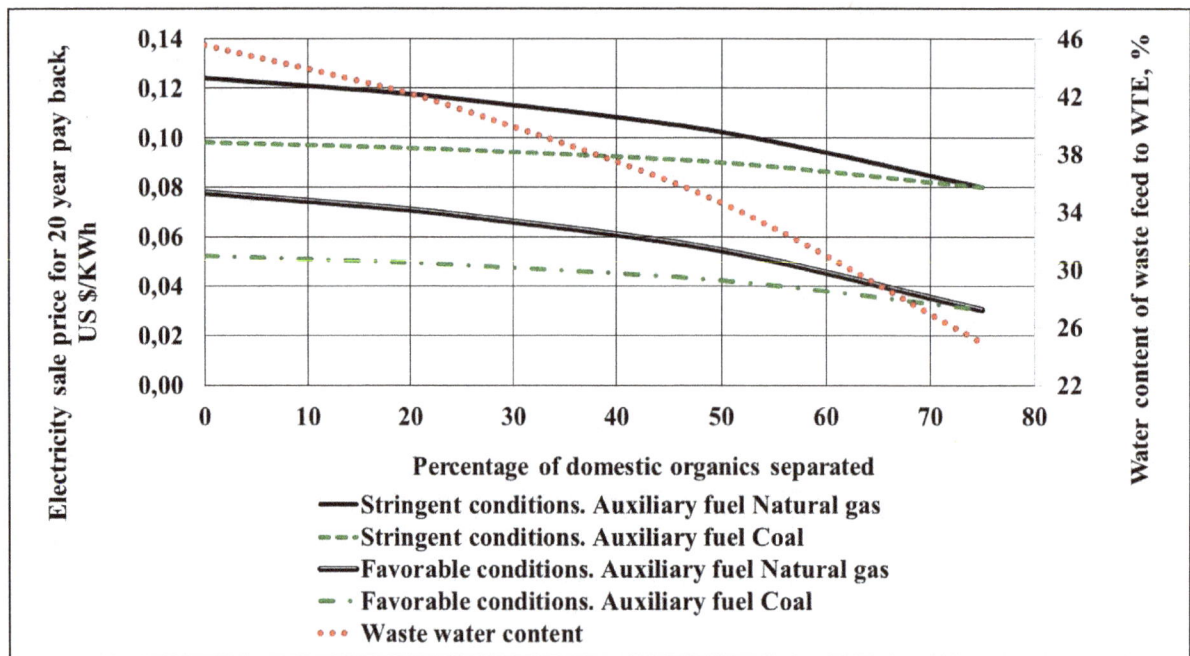

Fig. 4: Required electricity prices for an WTE cogenerating plant if a 20 year pay back is considered, based on yearly interest on debt of 6 and 10 %, applied to 50 and 100 % of investment (equity by owners) and gate fees of 0,010 and 0,020 US $/kg, both for natural gas and coal as auxiliary fuels

This figure shows the required electricity price for two situations: stringent and favorable. In them a 20 year-pay back is considered. Yearly interest on debt are either 6% or 10 %. Gate fee received are either 0.010 or 0.020 US $/kg (similar and twice the actual situation). In the stringent case, project owners do not contribute with equity (this means that the entire investment is to be financed by an external loan). In the more favorable case, owners contribute with 50 % of the investment as equity.

According to these simulations, the required electricity sale prices would be between 0.030 and 0.124 US $/kWh for the generation depending on the combination of conditions. Current electricity sale price available range is between 0.056 and 0.185 US $/kWh as sold to the final consumer. This means that there is room for finding an acceptable set of conditions.

The total number of 500 ton/day WTE plants in the country for this would be between 29 and 45. The total installed generating capacities would be between 400 and 850 MW. The estimated annual generated energy will be between 3500 and 7400 GWh.

The estimated investment for total treatment of separated waste by cogeneration WTE plants would be between 1600 and 3400 million US$. To have a perspective for these numbers consider that the local utility company, EPM is currently investing between 5500 and 6500 million US$ in the Hidroituango hydroelectric generating plant with an installed capacity of 2400 MW and 13930 GWh of annual generated energy. Another perspective comes from comparing the projected uses of natural gas and coal to the current yearly uses in the country, as seen in table 3.

Table 3: Comparison between expected use of cogeneration fuels in WTE system to generate electricity from waste in Colombia, for total treatment of separated waste by WTE

	Natural Gas	Coal
Fuel use in Colombia, ton/year	11 000 000	7 500 000
Average use for WTE for total separated cogeneration, ton/year	477 301	1 405 488
Average use for WTE for total separated cogeneration, % of total fuel use in Colombia	4.3	18.7

Putting these systems into functioning will require important proportions of the chosen auxiliary fuel usage in Colombia, especially in the case of the cogeneration with coal. However, in this case, Colombia is a large exporter and producer of coal for exports and there are enough reserves for this WTE application. Besides, some of the considered cases do not require auxiliary fuels.

After this simulation and analysis, it can be seen that it is necessary to add to the economic considerations the corresponding social and environmental evident and hidden benefits, which should be monetized as possible and included in the cash flow to evaluate a given project, comparing to landfilling and other alternatives. Among these, it is worth to mention the following ones:

- Job generation. This is especially important if a decision is taken to integrate, as much as possible, local engineering, installation and manufacturing into the project. This also applies to the jobs associated to the production of coal in the case of coal cogeneration.
- Contribution to eliminating greenhouse gases and to eliminating water and air pollution.
- Value and scarcity of the necessary land necessary for future landfills and expansion of existing ones and early recovery of valuable land of the existing landfills, for public use
- Cost of lixiviate and biogas treatment for any required landfill expansions.

- Valued added by the necessary research and development associated with the WTE systems and their technology. Intellectual property can be generated and negotiated.
- Transportation cost to far away landfills.
- Available electric power to equilibrate the offer when hydroelectric plants suffer for lack of rain and availability of electric energy to develop industrial and high technology agriculture in the sites close to the WTE plants.
- Possibility of uses for the ashes coming from the WTE plants.

Waste-managing concessionaires, responsible in great part for the final disposition of the MSW, should do this type of analysis. In so far, apparently, they have no seen yet WTE as an option to be considered.

The applicability of waste to energy system in Latin America has been studied and shown [2]. A previous general analysis of integral waste treatment, including WTE, has been proposed for the Medellin case [11]. It is important to find creative and innovative schemes in Colombia that allow promoting WTE projects [12,13], for example, sharing the initial investment between institutional actors and private actors or actors in the electricity sector. This means that the municipalities, regions and the national government could contribute with equity and investment in the stage of thermal or biochemical treatment and that the others participants invest from the turbine or internal combustion engine forward, for example. The public organizations could also provide steam, biogas or synthesis gas, which they could deliver to the private stockholders to produce and commercialize electricity and heat. To get some initial experience, it seems advisable,, to work out WTE projects, in Colombia with low electricity production, for example with capacities between 500 kW and 20 MW, practically the same range of small hydroelectric plants, and still small compared with the capacities of coal-fired thermoelectric and natural gas plants that exist in the Colombian electricity sector. However, each time a WTE project is executed that contributes with some MW power, it displaces the need for conventional large-capacity thermoelectric power plants over time, replacing existing ones or new projects, which in principle seem good from the macroeconomic point of view because coal and natural gas could be directed to sectors where their use is more efficient.

ACKNOWLEDGMENTS

The authors wish to thank Columbia University professor Nickolas J. Themelis, of the Department of Earth and Environmental Engineering; The City College of New York of the City University of New York professor Marco J. Castaldi. We also thank ACIEM Colombia and its Wtert Council for all the support and guidance.

REFERENCES

[1] P. Tello, E. Martínez, D. Daza, M. Soulier, H. Terraza. Informe de la Evaluación Regional del Manejo de Residuos Sólidos Urbanos en América Latina y el Caribe 2010. OPS-AIDIS-BID

[2] N. Themelis, M.E.Diaz, P. Estevez, M. Gaviota. Guía para la Recuperación de Energía y Materiales de Residuos. Original study sponsored by Banco Interamericano de Desarrollo. UNIVERSIDAD DEL DESAROLLO, CHILE; COLUMBIA UNIVERSITY, U.S.A. 2013. Edición en español, 2016

[3] Min Ambiente – Gobierno de Colombia – Noticias. 2016. A 2018 Colombia tendrá una tasa de reciclaje del 20%. Consulted on line in: http://www.minambiente.gov.co/index.php/noticias/2291-a-2018-colombia-tendra-una-tasa-de-reciclaje-del-20

[4] Documento CONPES 3874. CONSEJO NACIONAL DE POLÍTICA ECONÓMICA Y SOCIAL. REPÚBLICA DE COLOMBIA. DEPARTAMENTO NACIONAL DE PLANEACIÓN. POLÍTICA NACIONAL PARA LA GESTIÓN INTEGRAL DE RESIDUOS SÓLIDOS. 2016. Consulted on line in : https://colaboracion.dnp.gov.co/CDT/Conpes/Econ%C3%B3micos/3874.pdf

[5] Ministerio de Vivienda, Ciudad y Territorio. República de Colombia. Comisión de Regulación de Agua Potable y Saneamiento Básico. RESOLUCIÓN CRA 720 DE 2015. Consulted on line: http://www.metropol.gov.co/Residuos/Documents/

[6] REPÚBLICA DE COLOMBIA. Superintendencia de Servicios Públicos Domiciliarios. Informe Nacional de Aprovechamiento – 2016. Consulted on line in: http://www.andi.com.co/Uploads/22.%20Informa%20de%20Aprovechamiento%20187302.pdf

[7] IDEAM, Inventario nacional de gases de efecto invernadero GEI Colombia. 2015. Consulted on line in: http://documentacion.ideam.gov.co/openbiblio/bvirtual/023421/cartilla_INGEI.pdf

[8] OECD. Declaration on Environmental Policy. 14 November 1974 - C/M(74) 26/FINAL. Consulted on line in: https://legalinstruments.oecd.org/Instruments/

[9] DNP. Bases del Plan Nacional de Desarrollo de Colombia para el periodo del 2014- 2018. "Todos por un Nuevo País". 2014

[10] Ministerio de Vivienda, Ciudad y Territorio. DECRETO 1077 DE 2015. Diario Oficial No. 49.523 de 26 de mayo de 2015. Por medio del cual se expide el Decreto Único Reglamentario del Sector Vivienda, Ciudad y Territorio.

[11] Posada, Enrique. Strategic Analysis of Alternatives for Waste Management, chapter 2 in Waste Management (Edited by Sumil Kumar, ISBN 978-953-7619-84-8,

[12] Posada, Enrique. Reflexiones sobre el presente y el futuro de la ingeniería de proyectos. Dyna, año 79, Edicion Especial, pp. 14-16, Medellín, Octubre, 2012. ISSN 0012-7353

[13] Enrique Posada, 2017. "The culture of innovation and sustainable development: challenges for engineering", International Journal of Development Research, 7, (12), 17655-17660.

[14] WTERT- University of Columbia. http://www.seas.columbia.edu/earth/wtert/

CURRENT STATUS OF WASTE-TO-ENERGY IN CHINA

Binhang Hu, Zhejiang University

Qunxing Huang, Zhejiang University

Mujahid Ali, Zhejiang University

Yong Chi, Zhejiang University

Jianhua Yan, Zhejiang University

ABSTRACT

Innovative strategies are imperative to dispose of the growing municipal solid waste (MSW) through waste to energy (WTE) technology. The WTE progress in China deserves to receive particular attention as the rapid economic development with accelerated urbanization and industrialization leads to a tremendous amount of solid waste output, 203.62 million tons MSW in 2016. In recent years, the national and local government have both made tremendous effort to advance the development of waste-to-energy by issuing new waste management regulations, policies and WTE research projects. Based on these developments, an overview is needed to show the existing as well as new progress of WTE technologies in China. This paper summarizes the current status of solid waste including MSW and rural solid waste (RSW), introduces WTE capacity, several WTE incineration power plants and incinerator manufacturers in China. Besides, the paper shows some of the latest WTE progress including large capacity of domestic moving grate, high parameters of super-heated steam, ultra-low emission, mechanical biological treatment (MBT), new development in pyrolysis and gasification process, and fast continuous online dioxin monitoring.

INTRODUCTION

China, the largest developing country in the world, has the world's largest population of nearly 1.4 billion based on the population census in 2017. The country is experiencing a rapid economic development with accelerated urbanization and industrialization, which leads to a tremendous amount of solid waste output. In particular, municipal solid waste (MSW) collected in 660 cities had reached 203.62 million tons for 665.57 million people in 2016. Increasing MSW production has imposed huge pressure on the present landfills, and it also causes enormous damage to ecological environment. Therefore, the Chinese government is confronted with great challenges in dealing with these wastes [1]. Waste management should be based on the principles of reduction, reuse and recycling [2,3]. Based on the principles, Waste to Energy (WTE) conversion process refers to various technologies that recover energy from discarded waste including thermal conversion way (such as incineration, pyrolysis and gasification), biological treatment (such as anaerobic digestion and fermentation) and landfill [4]. The recovery products are mainly electricity, heat, steam and some combustible materials (such as char, bio-oil fuel, hydrogen, methane and other synthetic fuels) [5]. Thus, WTE is a prospective alternative to solve the waste production issue and a promising source of renewable energy (RE) [6,7].

Accordingly, WTE technology is currently in a stage of rapid development in China. A number of researchers have reviewed WTE application involving MSW collecting, recycling and disposal in various cities in China [8-12]. In recent years, WTE incineration power plant has been increasing dramatically, and started being gradually accepted as an important waste treatment method. In fact, MSW incineration technology can lead to a significant reduction of waste volume, prevent secondary gaseous pollution and provide electricity and heat for steam [13]. Based on the composition and recovery degree of some materials (such as metals) in ash, around 90 % of the total waste volume or 80-85 % of the solid waste can be reduced [14]. While there are also numerous difficulties and drawbacks in WTE incineration that have yet to be overcome. A number of emission pollutants, including nitrogen oxides, hydrogen chloride, sulfur dioxide, polycyclic aromatic hydrocarbon (PAH), heavy metals and dioxins, need to be treated in specially designed flue gas purification system before emitted in the atmosphere to meet the ultra-low emission standard recently proposed in China [15,16]. Also, the traditional off-line analysis method of dioxins can no longer serve the fast, online, real time purpose for dioxins detection [17,18]. In addition, the pretreatment of MSW in China should be paid extensive attention due to its particular characteristics of high-moisture content and unsorted complex components [19].

This paper reviews the current status of WTE technology in China and reveals some of the latest WTE progress, including large capacity of domestic moving grate, high parameters of super-heated steam, ultra-low emission, mechanical biological treatment (MBT), new development in pyrolysis and gasification process, fast continuous online dioxin monitoring.

WTE IN CHINA

Solid waste in China

Fig. 1 shows the trends in the amounts of MSW collected on the basis of 0.83 kg per person per day over the past decades. Apparently, MSW collected in 660 cities in China has reached 203.62 million tons for 665.57 million urban people in 2016. MSW collected has been augmenting at an average annual rate of 4.5 % from 2006 to 2014. While from 2014 to 2016, the average annual rate for MSW generation increased rapidly to 7 %, which results in a huge waste management problem in China. On the other hand, the rural solid waste (RSW) should also receive significant attention as there are still around 667 million people living in rural areas in China in 2016. Moreover, each person generated around 0.3 kg per day in rural area, leading to total 70 million tons RSW collected in 2016.

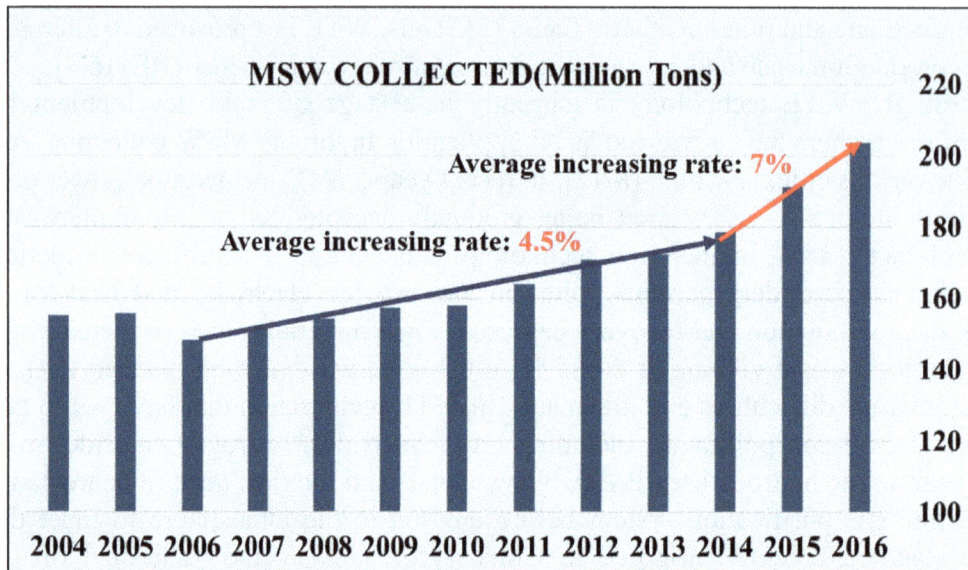

Fig. 1: The amounts of MSW collected from 2004 to 2016

Fig. 2 shows the amounts of MSW generation in 2016 and predicted data in 2025 and 2035 from different provinces in China in descending order. Due to the economic inequalities, there is a big difference of the MSW generation from different province. It reveals that the MSW production in Guangdong province ranks first, reaching nearly 25 million tons, followed by Shandong province. In addition, the amount of MSW generation in Zhejiang province reaches over 15 million tons, close to the data in Jiangsu province. By prediction, 439 million tons MSW will be produced for whole China, while the figure will increase to 547 million tons per year in 2035.

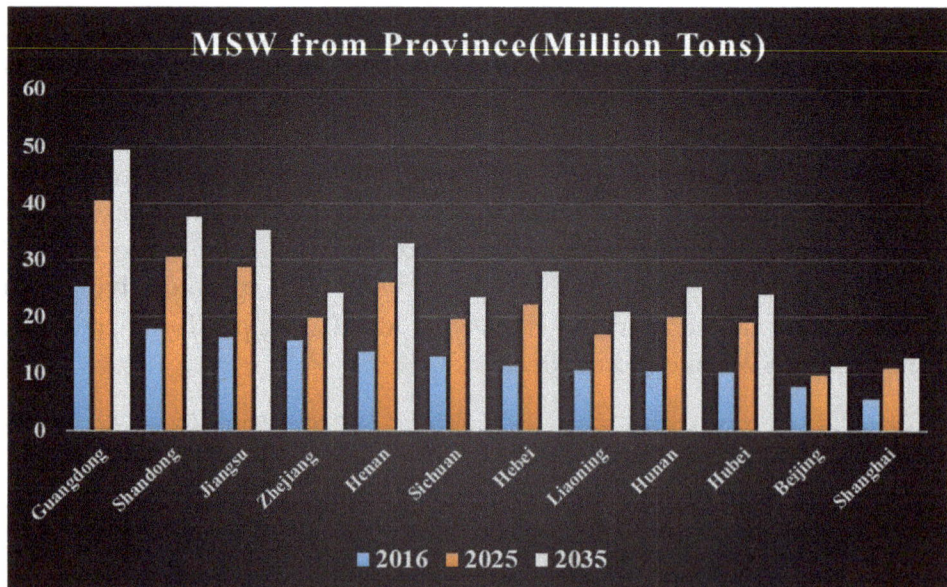

Fig. 2: The amounts of MSW collected from different provinces in China in 2016 (including predicted values for 2025 and 2035)

Capacity of waste-to-energy in China

Fig. 3 (a) and (b) reveal the WTE proportion of total waste generation and the amounts of WTE per day from 2004 to 2016, respectively. As shown in Fig. 3 (a), the WTE ratio increases steadily over time, reaching the maximum of 36 % in 2016. In detail, the amount of WTE achieves 202 kilotons per day in 2016 compared with only 12 kilotons 12 years ago. Fig. 4 reveals the prediction of WTE capacity in several provinces for 2025 and 2035 in China. According to statistics, the WTE capacity would reach 262 million tons per year at a ratio of 56% in 2025. While till 2035, the WTE capacity is expected to reach 408 million tons per year at a higher ratio of 71%. The maximum WTE capacity still locates in Guangdong province corresponding to a WTE capacity exceed 40 million tons per year in 2035. As for Zhejiang province, the WTE capacity would achieve around 16 million tons per year in 2025 and increase to 22 million tons per year in 2035.

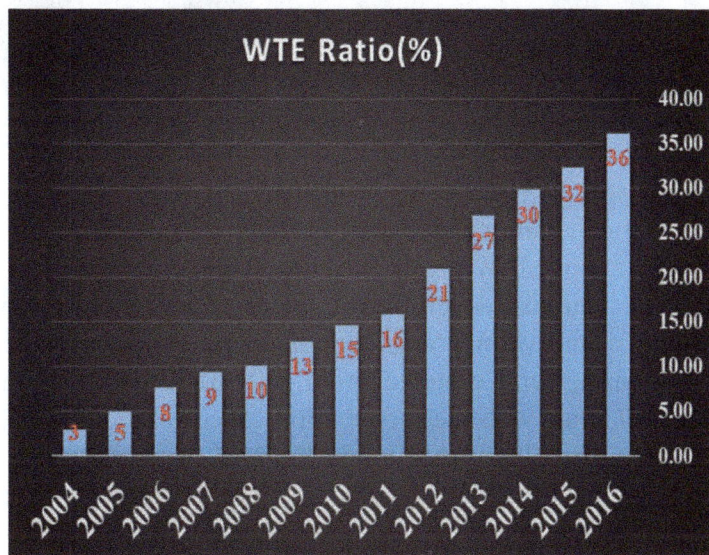

Fig. 3: (a) The WTE ratio of total waste generation

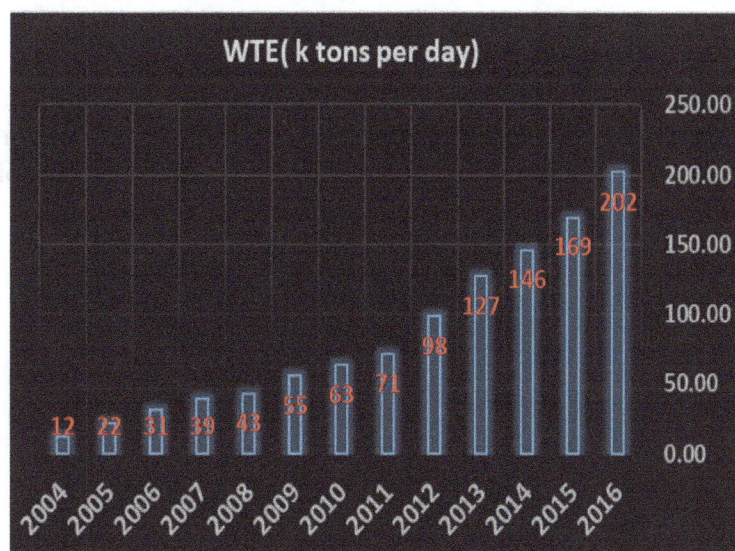

Fig. 3: (b) The amounts of WTE per day from 2004 to 2016

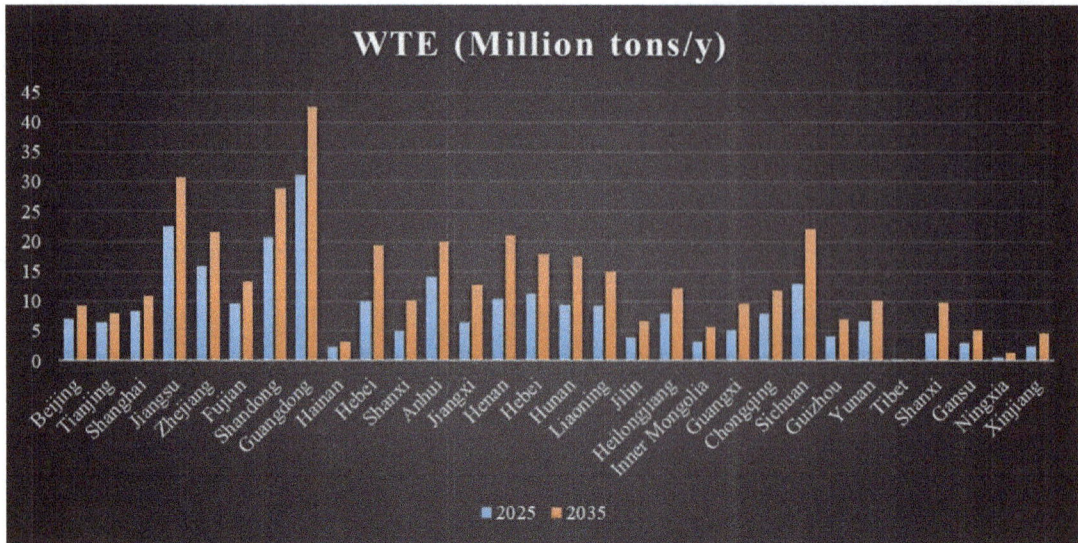

Fig. 4: The predicted WTE capacity in several provinces for 2025 and 2035 in China

MSW power generation capacity

The total MSW power generation capacity of is 7.251 million kilowatts, accounting for 49 % of the total installed capacity of MSW power generation, generating 37.514 billion kwh per year. From the perspective of regional development, there are great differences among different regions. As shown in Fig. 5, the MSW power generation construction speed of the eastern region is obviously ahead comparing with other regions. 180 MSW incineration facilities have been established in the eastern China, with the incineration capacity reaching 173,641 tons per day. A total of 46.989 million tons of MSW, accounting for 38.2 % of the total, are disposed meeting regulations limits through incineration. In contrast, only 46 MSW incineration facilities have been built in the central region, with the incineration capacity of 38,218 tons per day, out of which 11.595 million tons are domestic waste. In the western region, there are only 31 MSW power plants in operation, with an incineration capacity of 23,365 tons per day. The main way of disposing of domestic waste in western region is through landfill. Only 7,211 million tons, 16.5 % of domestic waste, are disposed by incineration. Nationwide, 70.0 % of the incineration facilities and 71.4 % of the non-hazardous disposal of waste incineration are concentrated in the eastern region. In general, 296 plants are in operation burning 102 million tons of MSW in 2017. Besides, there are also 120 plants under construction and more than 102 plants in planning in China. The average operating hours of waste incineration power generation achieve 5981 hours per year in 2017.

Fig. 5: The distribution map of MSW incineration power plants in China

Incinerators and WTE companies in China

According to the statistics of China Environmental Federation Research Institute, 250 domestic MSW incineration power plants had been put into operation, with a total capacity of 237 thousand tons per day and a total installed capacity of about 4880 MW by the end of 2016. Furthermore, there are 168 incineration power plants using grate furnace, with a total capacity of 164 thousand tons per day and installed capacity of 3040 MW. The rest are mainly fluidized bed incineration power plants, with a total quantity of 82 and a total capacity of 73 thousand tons per day and installed capacity of 1840 MW. Fig. 6 shows the quantity and proportion of grate furnace and fluidized bed used in MSW incineration plant from 2012 to 2016. Clearly, the proportion of grate furnace presents a steady upward trend, while that of fluidized bed had been decreasing year by year. In 2016, the fluidized bed proportion accounts for only 32.8 % with the quantity of 82. Specifically, a large number of MSW incineration projects had been completed and put into operation with an average treatment scale of 833 t/d and an average construction investment of 535,800 RMB/ton from 2015 to 2016, increasing dramatically from 2014.

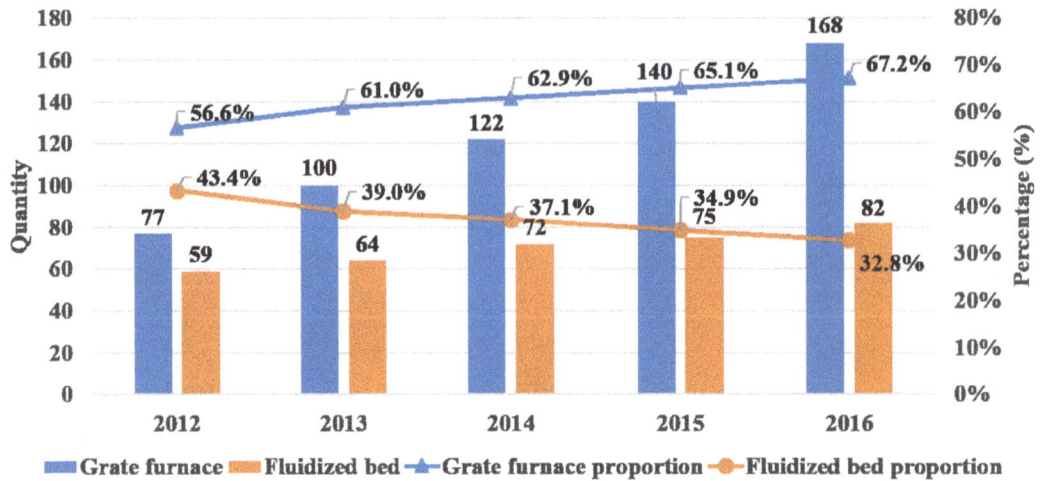

Fig. 6: The proportion and quantity of grate furnace and fluidized bed used
in MSW incineration plant from 2012 to 2016

By the end of 2016, the disposal scale of projects including planning in China's waste incineration market had exceeded 550 thousand tons per day. The top five companies are Everbright International, Jinjiang Environment, China National Environmental Protection Group, Chongqing Sanfeng Environment and Kangheng Environment, with market share of 11 %, 11 %, 9 %, 8 % and 7 %, respectively. The top five enterprises, account for over 40% of the waste incineration market. As for the waste incineration technologies, most grate stoke based incinerator are manufactured using technologies provided by European companies, such as MARTIN GmbH, B&W Volund, Steinmuller and Japan companies including Mitsubishi Heavy Industries, Hitachi Zosen Inova, JFE Engineering Corporation. While the Circulating fluidized bed (CFB) technology are mainly using technologies provided by domestic universities and research institutes, including Zhejiang University, Tshinghua university and the Chinese Academy of Sciences. The multi stage recirculating fluidized bed developed by Zhejiang University is mostly widely used and by using adiabatic furnace refractory wall, their CFB incinerator could burn high moisture MSW without auxiliary fuels and the largest capacity for a single line could reach 800 t/d.

WTE PROGRESS IN CHINA

Large capacity domestic MG

Due to the fast urbanization and limited land resources, big incinerator with large capacity is intensively required in near future for building WTE plants with massive capacity. There are already a number of plants with large capacity domestic MG in operation or planning in China. Established in 2017, Hangzhou Jiufeng waste incineration and power generation project adopts medium temperature and pressure steam parameters and is equipped with two steam turbines with rated output of 35 MW and two 40 MW generators. It provides a construction scale of 3000 tons of daily capacity and four moving grate incinerators with 750 tons of daily incineration of domestic waste. Also in Hangzhou city, another 5000 tons plant in already under construction. Now, the first domestic manufactured 850 t/d incinerator will be used in Everbright Yixing

Energy Co., Ltd, which can generate more than 600 kWh of electricity per ton of waste, equivalent to one month's electricity consumption of three households. Additionally, the MSW power plant project in Suzhou plans to reach a daily waste disposal capacity of 1800 tons equipped with two moving grate incinerators with 900 tons of daily incineration of domestic waste.

High parameters super-heated steam

Due to the high content of corrosive species contained in the MSW, such as chlorine, fluorine, sulfur and alkali metals, the temperature of combustion heat recovery super-heated steam is relatively low comparing to that of coal fired power plants to avoid high temperature corrosion of superheater. The steam parameters of most waste incineration power plants in China are set at 400 °C and 4.0 MPa. Under such parameter, the efficiency of the Rankine cycle is relative low and only 20% of the energy contained in the MSW is converted into electricity. As higher-parameter waste incineration technology can increase power generation, there is a trend of waste incineration technology to use high-parameter super-heated steam waste incineration technology for power generation. For instance, in 2014, Guangzhou Boluo waste incineration power plant increased steam pressure to 6.5 MPa at a temperature of 400 °C to enhance steam turbine efficiency. In June 2018, the first two-furnace-one-machine unit using high-parameter intermediate-reheating high-speed steam turbine combination technology of 6.5 MPa, 450 °C and 5000 rpm was successfully developed by Everbright International. It was built and put into operation in Jiangyin Phase III, with the cycle thermal efficiency reaching 26.5%. Meanwhile, the MSW power plant project in Suzhou is expected to use super-heated steam with medium temperature high pressure waste incineration power generation technology (13.7 MPa, 450 °C). As for high-parameter fluidized bed, Jinjiang Environment Shandong Zibo Linzi waste power plant adopts the first high-parameter circulating fluidized bed incinerator in China (7.9 MPa, 520 °C) with the treatment capacity of 2000 tons per day and the gross electricity efficiency reached 28%.

Ultra-low emission

In order to avoid possible secondary pollution caused by waste incineration, the implementation of ultra-low emission for MSW incineration power plant has gradually become a direction and goal in China. Zhejiang Ningbo Beilun waste to energy project is developed by Kangheng Environment with a daily disposal capacity of 1,500 tons, and the annual waste incineration power generation capacity is 199 million kWh, of which about 160 million kWh is integrated into the State Grid. The flue gas cleaning system consists of selective non-catalytic reduction (SNCR) to remove a part of NO_x at a high temperature, semi-dry scrubber injecting into Ca(OH)2 to react with acidic compounds effectively, activated carbon (AC) to adsorb a part of particulate matters, bag house to remove flying ash effectively, selective catalytic reduction (SCR) to effectively eliminate the rest NO_x at a low temperature, gas-gas heater (GGH) to increase the flue gas temperature and wet scrubber to further absorb acidic gaseous pollutants by alkaline liquids. The operation target value of this project is 5 mg/Nm^3 of SO_2, 50 mg/Nm^3 of NO_x, 5 mg/Nm^3 of HCl, 3 mg/Nm^3 of particulate matters and 0.01 ng-TEQ/Nm^3 of dioxins. Many other plants built in recent years are using similar APC (air pollution control) art to reach such ultra-low emission standard, such as Hangzhou Jiufeng WTE plant, under the requirement of local government.

MBT+CFB

Mechanical biological treatment (MBT) technology includes biological drying and mechanical separation process. The former process is a process of drying and dewatering preliminary crushing mixed waste by composting principle. Under the forced ventilation condition, micro-organisms ferment and produce heat by using perishable organic matter in mixed waste. Ventilation accelerates moisture volatilization at high temperature, and the moisture content of mixed waste decreases significantly, so as to achieve the effect of biological drying. After mechanical separation process, non-combustible materials, non-ferrous metals, plastics (PVC), glass (color classification) can be separated, so as to increase the recycling of resources. After sorting, waste can be incinerated directly and the efficiency of incineration power generation can be significantly improved.

After MBT process, the raw waste with heating value of 1000 kcal/kg and water content of 55-60 % would possess a heating value of about 2500 kcal/kg, and the water content would decrease to less than 30%. The waste in China contains a lot of organic matter and has a high moisture content. Combining mechanical biological treatment with waste incineration technology is the development direction in the future. In fact, the Shandong Zibo Linzi waste power plant with the treatment capacity of 2000 tons per day developed by Jinjiang Environment already adopts MBT to make solid recovered fuels (SRF) before incineration which can achieve a high gross electricity efficiency of 28%.

Pyrolysis and gasification

The present direct incineration methods of MSW are mainly in strong oxidizing environment, which can result in a greater risk of secondary pollution, especially the emission control of dioxins, heavy metals and other key pollutants concerned by the public. Thus, the construction of direct MSW incineration power plant is always resisted by the community of residents. Zhejiang University research group developed an advanced two-stage gasification combustion technology for domestic waste, combining with the present fluidized bed incineration technology. The front section adopts inclined reciprocating grate structure for preheating, drying and partial volatile gasification of domestic waste. The long rear arch is designed to lead the air flow to form a flue gas recirculation on the surface of MSW, meanwhile strengthening the surface radiation. Rotary grate and tertiary air are arranged at the tail, and the volatile matter and fixed carbon accumulate on the grate, forming a rotating upper exhaust gasification structure. This staged combustion technology adopts the subsection model of solid fuel gasification and gasification gas combustion. A large number of original pollutants in domestic waste remain in the solid phase, while only a small amount of restorative pollutants enters the gas phase in the pyrolysis and gasification process to participate in the follow-up reaction. Meanwhile, the formation of pollutants such as NO_x and dioxins can be avoided significantly due to the anoxic environment, and the dioxins contained in the original waste can be decomposed effectively in a high temperature combustion environment on the rear rotary grate.

It is difficult to achieve full combustion due to the complexity of hazardous waste (HW) composition and the great difference of component characteristics. Zhejiang University research team designed two upward and downward channels at the end of the rotary kiln, and developed a new multi-stage pyrolysis incineration technology which combines pyrolysis incineration in the kiln, gas phase component space combustion and solid residue spouting rotary distributor combustion. Unburned harmful gases and solid residues can enter the upper and lower channels respectively for secondary combustion. In this way, the capacity reduction after hazardous waste incineration has been greatly improved and the thermal reduction rate of incineration bottom slag has been reduced to less than 1%, reaching the international leading level.

Fast continuous dioxin monitoring

The dioxins emitted from waste incineration are one of the main sources of dioxins accumulated in the environment. At present, the common dioxin detection method around the world is the off-line analysis method of dioxin. Firstly, flue gas sampling (isokinetic sampling) is carried out on the spot, followed by sample pretreatment. Finally, high-resolution gas chromatography and high-resolution mass spectrometer (HRGC/HRMS) is used for ultima detection and analysis. Based on the study of correlation mechanism of indicators and the establishment of correlation model, Zhejiang University research team proposed a sampling module based on laser absorption spectrum tunable ionization-time-of-flight mass spectrometer (LASTI-TOFMS), coupled with constant-speed sampling method. and developed an incineration process II with independent intellectual property rights. An on-line rapid detection technology for dioxins in incineration process with independent intellectual property rights has been developed.

Specific implementation process of on-line rapid detection technology for dioxins in incineration process is presented. The flue gas is pumped into the sampling module by the sampling gun under the action of the sampling pump. After removing the particulate matter from the flue gas through the heating filter, it flows to the multi-way valve set, and then discharges to the atmosphere through the sampling pump. All the sampling pipelines and heating filters are heated to about 180 °C to prevent the target substances from condensing or adsorbing on the pipe wall. The water is removed from the flue gas (30-50 mL/min) by the Nafion drying pipe and then extracted to the pre-concentrator. Next, the organic matter in the flue gas is adsorbed and concentrated by the cold trap of the pre-concentrator, and then desorbed to the gas chromatograph for separation. The gas separated by gas chromatography enters the time-of-flight mass spectrometer through a pulse valve, ionized under the action of tunable laser, and then is detected by the time-of-flight mass spectrometer detector. The detector signal is obtained by the high-speed data acquisition card of the computer control module, and the dioxin toxicity equivalent concentration quantity is obtained through the calculation and analysis of the dioxin indicator correlation model.

THE PERSPECTIVE OF WTE IN CHINA

This section will provide a systematic outlook of the future of WTE in China. A system with functions of classification, collection and disposal will be established based on compulsory waste sorting. For instance, all domestic waste need to be classified into hazardous waste, perishable waste and recyclable waste in order to promote waste resource utilization. By the end of 2020, the relevant laws and regulations as well as standards system will be established as a replicable and propagable domestic waste classification model. At the same time, the recycling rate of domestic waste will reach over 35% in cities that implement mandatory classification of domestic waste. The government also will promote the resource utilization and non-hazardous treatment of kitchen waste. According to the amount and distribution of kitchen waste in various regions, the overall arrangements and scientific layout will be made as an entirety. In particular, the government will encourage the use of kitchen waste to produce oil, biogas, organic fertilizer, soil improver and feed additives. Furthermore, the treatment of kitchen waste with other organic degradable will be significantly promoted. At the end of the "Thirteenth Five-Year Plan" in China, the government will strive to increase the processing capacity of kitchen waste by 34,400 tons/day, and the system for recycling of kitchen waste will be mainly completed in cities. In addition, to further improve the energy utilization of waste, MSW incineration capacity will account for over 50% of the total capacity of non-hazardous treatment by the end of 2020, of which over 60% in the eastern region. The models including Public Private Partnership (PPP), franchising and environmental pollution third-party management will be vigorously promoted, as well as all types of social capital to actively participate in the investment, construction and operation of urban waste non-hazardous treatment facilities. In the process of promoting domestic waste reduction, resource utilization and non-hazardous treatment, science and technology innovation will be the main driving force.

CONCLUSIONS

Over the past decades, the Chinese central and local government have devoted significant effort to promoting the development of WTE technology in China. WTE incineration power plant has also been gradually accepted by Chinese people as a renewable clean method for recovering energy. WTE technology in China has also developed a number of innovative projects, including large capacity of domestic moving grate (2×900 tons per day for Suzhou project in planning), high parameters of super-heated steam (steam parameter for 13.7 MPa, 450 °C), ultra-low emission, MBT+CFB technology, combination of pyrolysis and gasification, fast continuous online dioxin monitoring. These progress of WTE technologies in China is expected to provide helpful references to other developing countries and even the whole world.

ACKNOWLEDGMENTS

The authors would like to greatly acknowledge National key research and development program [2018YFC1901300], National Natural Science Foundation of China [Grant No. 51621005] and the National Key Research and Development Program of China [2016YFE0202000], the Key Project for Strategic International Collaboration on Science and Technology Innovation of the National Key R&D Program of China [No. 2016YFE0202000], the Fundamental Research Funds for the Central Universities, Science and Technology Plan Project of Zhejiang Province [No. 2016C33005] and the Doctoral Graduate Academic Rising Star Program of Zhejiang University.

REFERENCES

[1] Cheng, Hefa, et al. "Municipal solid waste fueled power generation in China: a case study of waste-to-energy in Changchun city." *Environmental science & technology* 41.21 (2007): 7509-7515.

[2] Seadon, Jeffrey K. "Sustainable waste management systems." *Journal of Cleaner Production* 18.16-17 (2010): 1639-1651.

[3] Shekdar, Ashok V. "Sustainable solid waste management: an integrated approach for Asian countries." *Waste management* 29.4 (2009): 1438-1448.

[4] Tan, Sie Ting, et al. "Energy, economic and environmental (3E) analysis of waste-to-energy (WTE) strategies for municipal solid waste (MSW) management in Malaysia." *Energy Conversion and Management* 102 (2015): 111-120.

[5] Cheng, Hefa, and Yuanan Hu. "Municipal solid waste (MSW) as a renewable source of energy: Current and future practices in China." *Bioresource technology* 101.11 (2010): 3816-3824.

[6] Chen, Xudong, Yong Geng, and Tsuyoshi Fujita. "An overview of municipal solid waste management in China." *Waste management* 30.4 (2010): 716-724.

[7] Kalyani, Khanjan Ajaybhai, and Krishan K. Pandey. "Waste to energy status in India: A short review." *Renewable and sustainable energy reviews* 31 (2014): 113-120.

[8] Zhen-Shan, Li, et al. "Municipal solid waste management in Beijing City." *Waste management* 29.9 (2009): 2596-2599.

[9] Minghua, Zhu, et al. "Municipal solid waste management in Pudong new area, China." *Waste management* 29.3 (2009): 1227-1233.

[10] Zhao, Yan, et al. "Life-cycle assessment of the municipal solid waste management system in Hangzhou, China (EASEWASTE)." *Waste Management & Research* 27.4 (2009): 399-406.

[11] Zhang, Gang, Jing Hai, and Jiang Cheng. "Characterization and mass balance of dioxin from a large-scale municipal solid waste incinerator in China." *Waste management* 32.6 (2012): 1156-1162.

[12] Wang, Yuan, et al. "Effective approaches to reduce greenhouse gas emissions from waste to energy process: A China study." *Resources, Conservation and Recycling* 104 (2015): 103-108.

[13] Zhang, Dong Qing, Soon Keat Tan, and Richard M. Gersberg. "Municipal solid waste management in China: status, problems and challenges." *Journal of environmental management* 91.8 (2010): 1623-1633.

[14] Erik, A., et al. "Economic screening of renewable technologies: anaerobic digestion, and biodiesel as applied to waster scum." *Bioresour. Technol.* 222 (2016): 202e9.

[15] Vehlow, J. "Air pollution control systems in WtE units: an overview." *Waste Management* 37 (2015): 58-74.

[16] Hu, Yuanan, Hefa Cheng, and Shu Tao. "The growing importance of waste-to-energy (WTE) incineration in China's anthropogenic mercury emissions: Emission inventories and reduction strategies." *Renewable and Sustainable Energy Reviews* 97 (2018): 119-137.

[17] Hu, Binhang, et al. "Polychlorinated dibenzo-p-dioxins and dibenzofurans in a three-stage municipal solid waste gasifier." *Journal of Cleaner Production* (2019).

[18] Brunner, Paul H. "WTE: Thermal Waste Treatment for Sustainable Waste Management." *Recovery of Materials and Energy from Urban Wastes: A Volume in the Encyclopedia of Sustainability Science and Technology, Second Edition* (2019): 523-536.

[19] Zhou, Hui, et al. "An overview of characteristics of municipal solid waste fuel in China: Physical, chemical composition and heating value." *Renewable and Sustainable Energy Reviews* 36(2014):107-122.

"iENERGY-FROM-WASTE": EVOLUTION OR REVOLUTION IN AUTOMATION FOR MUNICIPAL WASTE TREATMENT FACILITIES?

Christophe Cord'homme, CNIM

INTRODUCTION

Municipal Solid Waste is an important sustainable source of material and a low-carbon renewable energy. Material-from-Waste and Energy-from-Waste (EfW) plants contribute to the diversion of biodegradable municipal waste from landfills and secure a noteworthy reduction in greenhouse gas emissions.

Automation and information technologies are at the forefront of innovation in our day-to-day lives, but also in industrial sectors. The purpose of this paper is to understand the activities and projects of CONSTRUCTIONS INDUSTRIELLES DE LA MÉDITERRANÉE (CNIM) regarding this trending topic; sustainable waste management. CNIM is an international integrated company supplying turnkey Design & Build (EPC) supplemented by Operations & Maintenance services for municipal solid waste treatment facilities. CNIM is designing state-of-the-art automation systems for Waste-to-Energy and waste sorting facilities and is proposing innovations, which are adapted to this specific industrial sector.

A TRENDING TOPIC

Computer sizes, speeds and memory capacities have improved over the last 50 years from kilo-units to Tera-units (octets, Hz). Today, these characteristics have been multiplied by approximately ten to the ten (1010).

With the help of this huge advancement in Information Technologies (I.T.), we are now talking about automation, digitization, the Internet of things, big data, robots and artificial intelligence in our day-to-day lives.

Of course, these concepts are also trending topics in most industrial sectors. This paper intends to explain the impact of IT for waste treatment and to show which innovations an industrial technology supplier such as CNIM could offer.

ARE ROBOTS FOR WASTE DISPOSAL AVAILABLE?
IS IMUNICIPAL-SOLID-WASTE OUR FUTURE?

The topic of robots for waste was the main subject of the movie "Wall E". The 2008 Sci-Fi animation film was already dealing with the day-to-day "life" of a robot dedicated to waste disposal. This robot, named Wall E for "Waste Allocation Load Lifter-Earthclass", could be considered at first sight to be the Omega of the "perfect" future of the automation of waste management. According to me, this looks more like a "hell" for the humanity and a dead-end for the consuming society.

CONTROL ROOMS, VISIBLE EVOLUTION

More seriously, Energy-from-Waste and Material-from-Waste facilities are using large industrial complex process. They require automation for their control. A long and visible evolution of the interfaces has occurred in recent decades. For example, XXIst century control rooms have changed a lot from XXth century ones.

Fig. 1: Paris EfW control room 1969

Fig. 2: Nantes (France) EfW control room 1987

Fig. 3: Oxford (UK) EfW control room 2014

Fig. 4: Kemsley (UK) EfW control room 2019

The above figures show the evolution with some photos of CNIM EfW control rooms, from Paris EfW 50 years ago to the latest design done for the Kemsley EfW plant starting in 2019, which has been designed with very detailed ergonomics, lay out and lighting studies. In between these are the designs used in the 1980's and the 2010's.

We are now even in position to detach process information from the control room and transfer it to the site. Augmented/virtual reality glasses are now being considered for this purpose.

PROCESS CONTROL CONDITIONS

I.T. is improving the way we understand and control processes, the way we collect and monitor measurement data from instruments and react to move actuators and motors. Nevertheless, this IT approach is not changing the physical principles themselves: processes, instrumentation & actuators are not virtual. They are there on site to deal with real and dirty material: waste!

An integrated approach is an approach using both technologies and IT to influence the design, building and operation & maintenance of waste facilities.

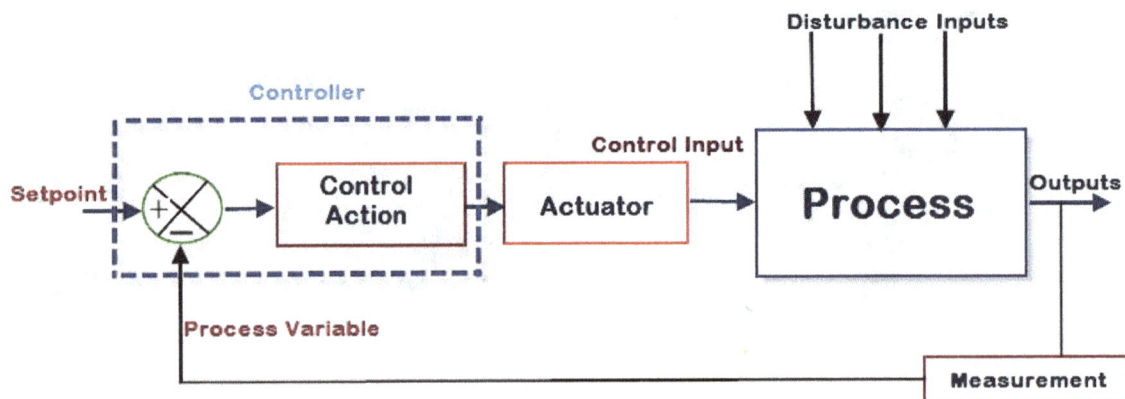

Fig. 5: Process control diagram

For this "delicate" automation technology dealing with "waste" treatment conditions, we need to increase:

- Flexibility,
- Availability
- Reliability,
- Safety under difficult conditions for better process controls, better performance and emissions monitoring at lower costs. This is why pragmatic integrated solutions are required.

ADVANCED AUTOMATION PROCESS CONTROL

Several parts of the plants are showing accelerations in automation and data management, such as combustion logics, smart energy recovery, Predictive Model Control for flue gas treatment or waste pit management. For example, the various EfW cranes' operation modes could be manual, semi-automatic for feeding, automatic for feeding, and now even unmanned full automation dealing with feeding, stacking (receiving), mixing and recasting. These can be operated either from a touch screen HMI panel in manual and semi-automatic modes from the standard operator's seat with a direct view over the waste pit or, more recently, through a Main User Interface (MUI) in un-manned full-automation mode. This remote system could be deported to a control room without any direct view over the pit using several video cameras.

For example, for the SNCR (Selective Non Catalytic Reduction) deNOx control process, DCS progress has improved things based on a changeover from Proportional-Integral-Derivative Control function (PID) to SCNR+ with Model Predictive Control (MPC). The urea set point is calculated as an advanced function based on NOx and NH_3 emissions. This control improvement could be compared to the principle of the control of an autonomous vehicle.

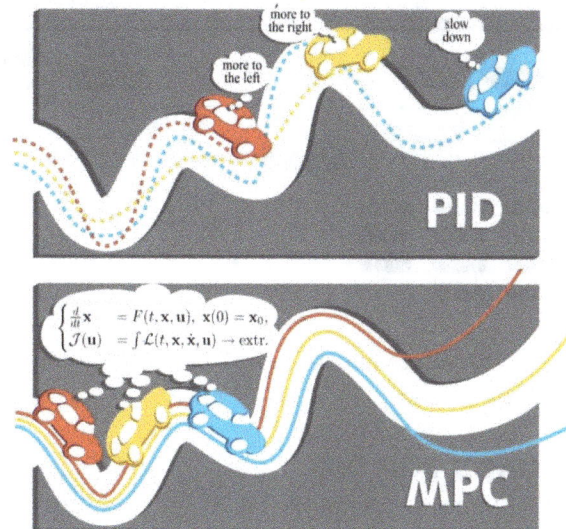

Fig. 6: PID control vs Model Predictive Control

DIGITAL CONTROL SYSTEMS (DCS)

Very powerful Digital Control Systems (DCS) have been developed to measure, calculate and act on the process. Safety system logic sequences are also programmed to protect the operation.

Typically, between 10,000 and 20,000 signals are controlled in an EfW plant, with hundreds of kilometers of wires travelling around the plant. The very high availability required by the plant (24/7/365) means that big data management needs to be done with high reliability. This means redundancies, remote access to the various package units of the plant, but with traceability and cybersecurity based on firewalls and controlled limited access.

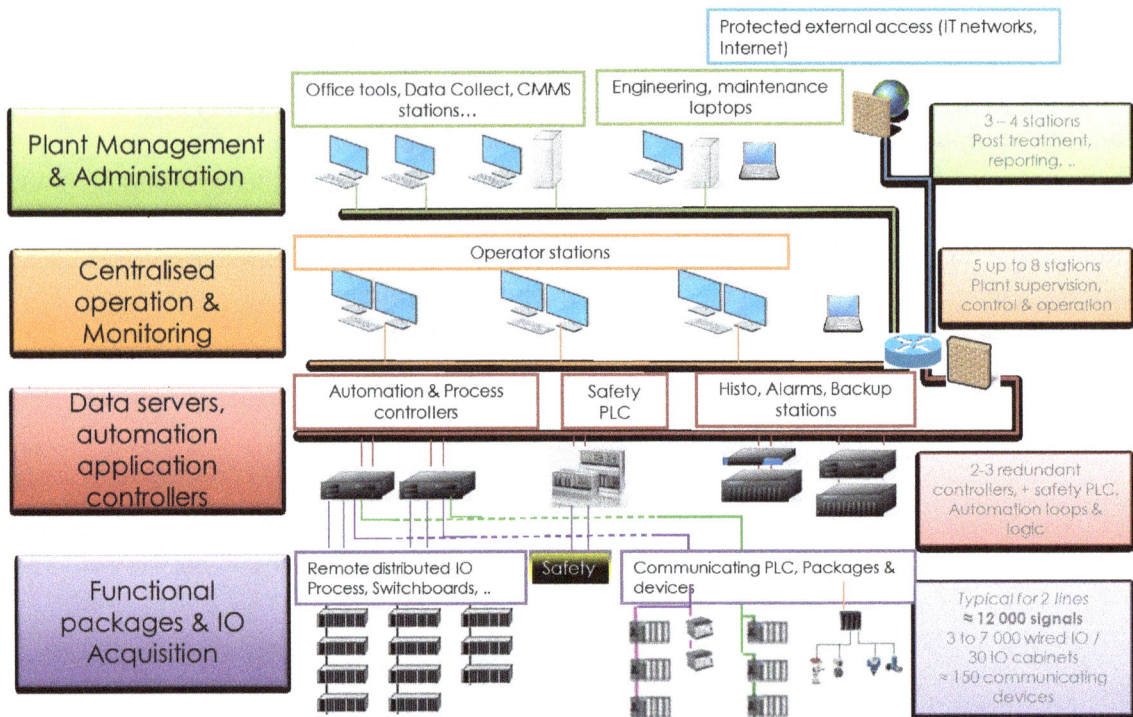

Fig. 7: typical advanced CNIM EfW architecture

CNIM DEDICATED APPROACH FOR DCS

The objectives of these advanced complex DCS architectures are to achieve optimum reliability, as well as cyber-security with big data management and remote access. That is why CNIM has decided to develop the entire application internally, considering the critical aspect of the DCS procedure and structure, the importance of the DCS interfaces for all the various processes in the plant and the safety requirements! This is the "brain" of the plant!

The fields of action of the CNIM automation solutions department are mainly:

- The development of Digital Control Systems in collaboration with DCS manufacturers
  - Programming of DCS logic
  - Implementation of the plant's views in the DCS
  - Programming of safety system logic
  - Factory Acceptance Tests
- Commissioning of Digital Control Systems
- Technical support for customers
  - Remote assistance (more than 25 sites are now connected to CNIM's hub).
  - On-site preventive maintenance
  - Training for operations and maintenance staff
  - DCS upgrades
- Computer Assisted Maintenance Management System

Digitization has also improved reporting and data management. Operators can obtain remote technical assistance from experts. Process simulation is also available now. It is enabling on-line training for operations and maintenance staff. One should not forget that on-site connectivity also requires cyber security measures for its development.

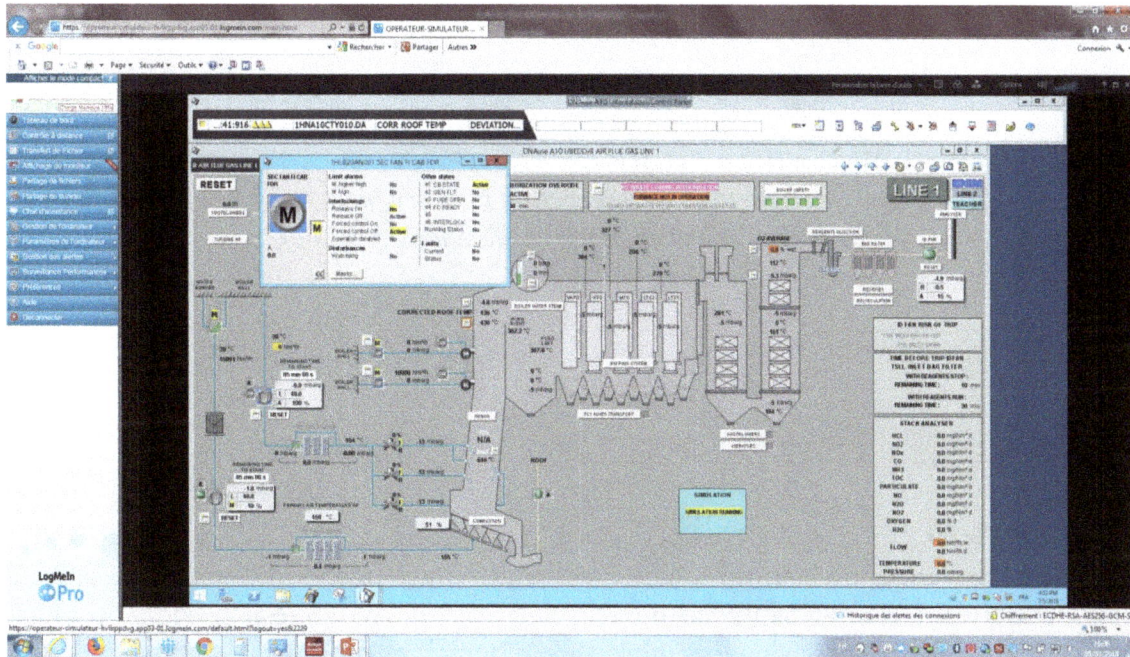

Fig. 8: CNIM process simulation available on remote access for training

Lastly, Computer Assisted Production/Maintenance Management Systems are also improving flexibility, availability, reliability and the safety of these waste treatment facilities.

One should not forget that advanced process controls, advanced performance and emissions monitoring should be done at the lowest possible cost, considering the difficult technical conditions of waste management. It should lead us to choose pragmatic solutions.

CONCLUSIONS

Automation in waste management is not a new topic, but one that is in constant evolution with the help of IT technical advancements! The CNIM automation solution presents a reliable, state-of-the-art, comprehensive and pragmatic automation solution for waste management plants. The modern DCS is designed internally with full redundancy and security, and benefits from CNIM's experience from the design stage to construction and operation. The integrated process control and safety systems are improving performance monitoring solutions, remote access using industrial Internet and process simulations. With this in-house technology, automation design and construction work can integrate operational tools such as Computer Assisted Production and Maintenance Management, Integrated Logistic Support (ILS) and Reliability, Availability, Maintainability & Safety (RAMS). This illustrates the rapid evolution of automation in Resources-from-Waste facilities. No revolution of "iMUNICIPAL-SOLID-WASTE" or the 'Internet of Waste' (equivalent to the "Internet of Things"), but "iEfW", ("iEnergy-from-Waste") is already available!

BUILDING THE FUTURE OF WASTE AND RECYCLING
Changes within the waste and recycling industry have sparked ideas
for how to better manage waste and recyclables.

Mallory Szczepanski, Waste360   Cristina Commendatore, Waste360

INTRODUCTION

The waste and recycling industry is currently undergoing a number of changes, mostly due to China's import restrictions and a fluctuating recycling market. And in response to those changes, members of the industry are having to rethink some of their waste management processes.

At the 2018 EEC/WTERT Bi-Annual Conference, which was held October 4 and 5 at the Earth Engineering Center (EEC) at the Grove School of Engineering, City College of New York, industry experts gathered together to discuss waste-to-energy (WTE), the concept of a circular economy, end-of-life product design and potential opportunities to help combat contamination and improve safety across the industry.

Here's a deep dive into some of the topics discussed.

ADAPTING TO CHINA'S NEW WASTE IMPORT REGULATIONS

China, the largest importer of post-consumer recyclables, implemented a waste import ban on 24 kinds of solid wastes in January and a new contamination standard of 0.5 percent in March 2018. These restrictions, along with more coming down the pike, were put into place in an effort to stop "foreign garbage" from inundating the country and have since caused the U.S. recycling industry to scramble to find both short-term and long-term solutions for managing recyclables [1,2].

For some time, China has been receiving low-quality and contaminated materials from the U.S. and others. But now, with these new restrictions and limited end markets, the global recycling industry is being pushed to actually make improvements and to generate clean, high-quality materials.

In an effort to clean up the recycling stream, haulers, recyclers and municipalities are making changes to single stream recycling programs, utilizing more technology, ramping up education, etc. And industry associations, including the National Waste & Recycling Association (NWRA), the Solid Waste Association of North America (SWANA) and the Institute of Scrap Recycling Industries (ISRI), are stepping in to help by communicating directly with China and encouraging members of the industry to take action to improve recycling and reduce contamination.

"NWRA has always supported China's efforts to improve its environment. However, we believe there are better ways to achieve those goals than to tighten restrictions on imported recyclables," said Darrell Smith, president and CEO of NWRA. "We have said before that the 0.5 percent standard would be nearly impossible for our members to meet, and it could cause some short-term disruptions to the industry. However, it could also present opportunities as our members continue to adjust."

"In the short term, we need to prioritize education, update technology in materials recovery facilities (MRFs), identify new markets for materials and be more transparent about costs associated with recycling as well as the revenue that's generated from the sale of recyclable materials," said David Biderman, executive director and CEO of SWANA.

Following through with these efforts, along with bringing new ideas to the table, will further help the industry prepare for the next wave of restrictions. Beginning December 31, 2018, 16 more types of solid waste, including compressed car scraps and scrapped ships, were banned from import. And beginning December 31, 2019, another 16 types, including stainless-steel scraps, will be banned.

"The Chinese government's announcement will have an impact on more than 676,000 metric tons, worth about $278 million, in U.S. scrap commodity exports to China in the first year and another 85,000 metric tons worth more than $117 million in the second year," said Robin Wiener, president of ISRI. "Although we anticipated more import restrictions would be announced, we remain concerned about the effect these policies have on the global supply chain of environmentally friendly, energy-saving scrap commodities and will instead promote an increased use of virgin materials in China, offsetting the government's intent to protect the environment."

In addition to those expanded restrictions, China is now considering a total import ban [3], which could go into effect by 2020 or 2021. (*Resource Recycling*'s timeline [4] has the most up-to-date information on the moves coming out of China.)

THE STATE OF WTE IN CHINA AND THE U.S.

Waste-to-energy, the process of generating energy in the form of electricity or heat from the treatment of waste or the processing of waste into a fuel source, is a method commonly used in areas like Europe and China. However, the U.S. and Canada have made strides in building and operating more WTE facilities over the years [5].

According to Bruce Howie, P.E., vice president and professional associate at HDR Inc., in the 1950s and 1960s, construction of early WTE facilities began in Europe and construction of incinerators started in the U.S. In 1975, construction began on the first WTE facility in North America, which was located in Saugus, Mass. The 1980s were the "hay days" of WTE in North America, and the 1990s were the "dark ages," because there were stricter emissions standards in the U.S. that led to retrofits and some facility closures. The 2000s brought expansions to existing facilities and construction of new facilities in Europe and Asia. Today, there is some development of new facilities in the U.S. and elsewhere, but tough economics are leading to some closures.

In the current economy, there are many challenges to overcome in the U.S., including there's no regulatory or policy framework currently in place to incentivize programs, there's limited to no funding or support to look beyond the traditional linear business model and there's no compromise between parties and stakeholders on either side of the aisle.

That said, Howie noted the key drivers of the expansion of WTE include a push for greater diversion from landfill, an enhancement of the "3Rs" (reduce, reduce and recycle) programs and organics collection programs, the idea of waste is a resource (the fourth R) and an interest in alternatives to traditional WTE technologies like mixed waste processing and anaerobic digestion.

While those key drivers are "doable," there are some key inhibitors to the expansion of WTE, including the fact that it's cheap to landfill materials, energy prices and fuel prices are relatively inexpensive (at least in the U.S.), there are many groups who oppose the idea of WTE and there are a number of regulatory factors to consider, such as carbon taxes and greenhouse gas legislation.

In China, the state of WTE looks much different. According to Prof. Qunxing Huang of Zhejiang University, from 2006 to 2012, municipal solid waste (MSW) in China increased at an average rate of 4.5 percent, and from 2014 to 2017, MSW in China increased at an average rate of 7 percent.

"[Looking forward,] we predict that in 2025, we will manage 429 tons of MSW per year for all of China, and in 2035, we will manage 547 million tons of MSW per year for all of China," he stated during the 2018 EEC/WTERT Bi-Annual Conference.

As of 2017, China has 296 WTE plants in operation, burning 102 million tons of MSW every year. The country now has another 120 plants under construction and 102 more in planning.

While China continues to explore new opportunities for WTE, the U.S. is beginning to discover its own opportunities for managing waste and recyclables. One standout opportunity is technology, which is being used more and more in WTE facilities, on fleets and in MRFs.

TECHNOLOGIES TO COMBAT CONTAMINATION AT MRFs

One of the biggest challenges in the fiber end of the market right now is contamination. And that problem has become an even greater concern over the last 12 months amid the implementation of China's waste import restrictions [1].

China has prohibited contaminated exports coming out of MRFs, and in response, other countries in Southeast Asia like Thailand [6] and Vietnam [7] have started to implement their own import bans due to their own capacity problems. As a result, MRFs and manufacturers alike have been looking to advanced developments in technology to improve the materials processed at their facilities, particularly when it comes to source separation and the quality of paper coming out of residential, single stream MRFs [8].

Some facilities have implemented high-speed optical sorters, ballistic separators and non-wrapping screens. Many facilities are also updating their older, antiquated equipment to properly separate and clean up what comes out of MRFs. While others are using artificial intelligence (AI)—robots—or a combination of technologies [9]. But the consensus throughout the industry is that something needs to be done from a policy and public education standpoint before materials even make their way to the MRFs.

In order to remove contaminants, companies like CP Group and Van Dyk Recycling Solutions have looked to a newer generation of optical sorters. CP Group's MSS produces optical sorters that run belts at 1,000 feet per minute (fpm) rather than 500 fpm, which is what the company does on the container side. The MSS FiberMAX is what the manufacturer released as its next-gen sorter, and it's a fairly intricate system that uses a near infrared spectrometer technology that allows the machine to decipher different materials from each other [10].

Another problem at MRFs occurs when plastic bags go into the same stream and mix with all the other materials. Van Dyk developed a screen, called a 440 Non-wrapping Screen, that film bags don't wrap on. Regardless of what it's fed, the technology will screen by size and angle at the same time.

Manufacturer Bulk Handling Systems (BHS) has developed its Max-AI technology, which the company claims is changing what is possible with MRF design and performance. The technology allows MRFs to process better-quality recyclables at a lower cost per ton. BHS also is using detection technology for optical sorters and MRF intelligence [11].

Machinex Technologies Inc., which designs and manufactures single stream recycling systems and machinery for MRFs, has taken a multipronged approach on how the company ought to address the fiber side of the business and has made new developments to its old corrugated containers (OCC) screen. On the sorting side of operations, Machinex sees the benefits of MRFs using optical units in combination with manual sorters to remove the brown and light-weighted/flattened containers from the fiber.

In order to complement some of the technologies companies are already using, some are turning to robotics to help sorting operations run more efficiently. And amid a growing labor shortage, MRFs are finding robotics could be beneficial [12].

In addition, infrared technology combined with color sensors and metal detectors in the high-speed sorters have been key to MSS' offerings. MRFs can add these optical sorters on each line and cut workers from the sort line, which could be considered a bonus amid an industrywide labor shortage. Machinex's SamurAI is designed to help minimize labor and set new goals for purity and efficiency. Right now, SamurAI is only available on the container side [13].

At the end of the day, MRFs and technology providers believe the solution to handling contamination at MRFs will most likely be a combination of technologies, as well as public education and regulation.

EDUCATION, PUBLIC OUTREACH AND REGULATIONS

When it comes to the education aspect, MRFs and their manufacturers aren't the only ones who cite the importance of public awareness and education. Paul Davison, the managing director of Proteus Communications Group in the U.K., stressed the importance of the public perception and not using phrases like "circular economy" when dealing with the public.

"They'll have no idea what you're talking about, and if you use jargon like that, they will stop listening to you straight away, because that's an exclusive language," said Davison during the 2018 EEC/WTERT Bi-Annual Conference. "They will think you are trying to be better than they are and that you are trying to con them. We have to watch the language we use when we're talking to the public."

According to Davison, there are some simple processes that can be used to educate the public. He stressed that when talking to the public about WTE, generally speaking, the public will think of a facility that emits noxious fumes and will likely have a dated view of the way technology works. Ultimately, people will think filth and dirt.

But realistically, he pointed out, the industry has to better drive home the benefits of advanced technology and getting value out of something that is actually waste.

"In Denmark, when I am talking to people about energy-from-waste, I don't have to explain to them the advantages. They know it. And we need to share that with other communities that don't know it," explained Davison. "I generally believe that the United States has a great opportunity to replace some of the filthiest, decrepit and out-of-date energy-from-waste plants I've seen on the planet. You've got to clean up those energy-from-waste plants. This technology is clean and has a purpose, and its role in a circular economy is essential. If you just show the public the dirty technology that is here in America, you will reinforce that image to the public, and it will make your job engaging the public that much harder."

Davison pointed to three disconnects between the public and waste:

- *Disconnect 1:* People do not view waste as their problem.
- *Disconnect 2:* Over the last 10 to 15 years, the focus on recycling has isolated residual waste disposal.
- *Disconnect 3:* Waste and resources are not the same thing.

In an effort to remove plastic bags and other non-recyclable items from the stream, various municipalities around the country are trying out programs to study and improve residential recycling habits.

Places like Massachusetts, Atlanta, Chicago and Orange County, Fla., have launched and studied programs that involve recycling experts looking inside curbside carts and leaving tags with recycling feedback [14]. For instance, carts are given weekly tags ranging from Great Job (near perfect), Good Try (a few recommended tweaks) or Oops (needs improvement) with some specific guidance to explain what specifically needs improvement.

In addition, since China's waste import ban went into effect in January 2018, technology companies launched effective and efficient waste and recycling apps and food waste reduction apps to help the public reduce contamination and recycle more efficiently [15]. These apps help users manage their waste and recycling, find nearby pickup or drop-off locations, rescue food that would otherwise go to waste and increase their knowledge about what can and cannot be recycled.

THE PUSH FOR A CIRCULAR ECONOMY

The concept of a circular economy has been around for some time now, but over the years, companies, organizations and individuals have stepped up to the plate to find new ways to reuse materials and design products with the end of life in mind.

According to Henrietta Goddard, a research analyst at the Ellen MacArthur Foundation, the main principles of a circular economy include: thinking about the impacts of the materials from the very beginning of the design process so that waste and pollution can be significantly reduced and designing products to stay in use and keep circulating to regenerate natural systems.

Those principles have been adopted by some manufacturers, designers, architects and others in an effort to make products last longer, reduce waste and promote the concept of reuse.

In Michigan, two leading furniture companies, Herman Miller [16] and Steelcase [17], have taken the lead on designing furniture and products with end of life in mind, as well as developing effective circular economy models [18].

For more than a decade, Steelcase has been piloting a number of different circular business models, focusing on designing products for circularity, finding new outlets for usable furniture, diverting food waste from landfill and more.

In addition, the company has created two end-of-use programs available to customers around the globe. Eco-Services, which services Europe, is a program that helps companies evaluate their furniture inventory and identify options for reuse, donation and recycling. And Phase 2 is the U.S. equivalent of that program with slightly different offerings.

Over the years, the two programs have been a success when it comes to keeping furniture out of landfill. In 2016, for example, Eco-Services handled 30,600 cubic meters of material, and Phase 2 diverted 6.4 million pounds of furniture from landfill from more than 1,300 decommission requests.

"In our industry, design trends are changing daily, and we need to be able to change as the industry changes," says Dan Dicks, director of global end-of-use services at Steelcase. "As a company, we have to be with our customers throughout their entire journey, essentially helping to eliminate the model of take, make and dispose. We plan on continuing to share our circular economy approach with our customers to achieve this and create even more products and services that will benefit the environment, the customer and Steelcase."

In an effort to reduce furniture waste, Herman Miller launched its rePurpose Program about nine years ago, which deals with surplus corporate assets through a combination of resale, recycling and donation. The program, which is managed by environmental firm Green Standards, helps companies meet sustainability goals and aid nonprofits by providing them with items they may normally have a hard time getting, such as large amounts of desks and chairs.

Over the years, Herman Miller has ramped up its sustainability efforts by creating products that are considered cradle-to-cradle in terms of the materials they are made with. Additionally, the company has worked to make its products easy to disassemble, ultimately giving the products a longer lifespan.

Recently, brands like Danone [19], SC Johnson [20], PepsiCo [21], Unilever [22] and others have made commitments to close the loop and reduce waste via recycling and reuse. These efforts are growing and being adopted by more and more brands across the globe, as more companies aim to be sustainable and more customers look for brands that are "going green."

CONCLUSION

Since China enacted National Sword, the industry has indeed been scrambling to find efficient ways to reduce contamination and strengthen the quality of materials coming out of MRFs. However, rather than viewing these stringent regulations as a burden, we should focus on how we can benefit domestically. This is our chance for businesses, designers, legislators, innovators, waste generators, residents and the industry as a whole to combine efforts to develop efficient, profitable and environmentally sustainable ways to better manage waste.

REFERENCES
[1] https://www.waste360.com/white-papers/chinas-changing-import-regulations-what-does-it-all-mean
[2] https://www.waste360.com/legislation-regulation/industry-associations-respond-china-s-expanded-waste-import-ban
[3] https://www.waste360.com/legislation-regulation/will-china-delay-total-import-ban-until-2021
[4] https://resource-recycling.com/recycling/2018/02/13/green-fence-red-alert-china-timeline/
[5] https://www.waste360.com/waste-energy/state-waste-energy-us
[6] https://www.waste360.com/legislation-regulation/thailand-ban-foreign-scrap-plastic-imports

[7] https://www.waste360.com/legislation-regulation/vietnam-malaysia-set-waste-import-restrictions

[8] https://www.waste360.com/mrfs/tackling-contamination-era-e-commerce-china-s-import-ban

[9] https://www.waste360.com/mrfs/what-robotics-and-ai-could-mean-future-industry-part-one

[10] https://www.waste360.com/mrfs/mss-unveils-new-optical-fiber-sorting-technology

[11] https://www.waste360.com/equipment/bhs-expands-max-ai-aqc-product-line-aqc-2

[12] https://www.waste360.com/labor-relations/truck-driver-shortage-projected-hit-all-time-high-2017

[13] https://www.waste360.com/equipment/machinex-debut-samurai-sorting-robot-wasteexpo

[14] https://www.waste360.com/residential/orange-county-fla-reveals-key-data-recycling-pilot

[15] https://www.waste360.com/fleets-technology/seven-handy-waste-and-recycling-apps

[16] https://www.waste360.com/waste-reduction/general-motors-herman-miller-and-green-standards-partner-landfill-free-solution

[17] https://www.waste360.com/waste-reduction/how-steelcase-s-initiatives-support-circular-economy

[18] https://www.waste360.com/waste-reduction/how-companies-are-addressing-issue-f-waste

[19] https://www.waste360.com/recycling/danone-accelerates-transition-toward-circular-economy

[20] https://www.waste360.com/plastics/sc-johnson-boosts-plastic-recycling-and-reuse-commitments

[21] https://www.waste360.com/recycling/pepsico-unveils-new-packaging-goal-25-recycled-plastic-2025

[22] https://www.waste360.com/plastics/unilever-and-veolia-collaborate-sustainable-packaging